Palgrave Studies in Disaster Anthropology

Series Editors
Pamela J. Stewart
Academic/Ed
University of Pittsburgh
Pittsburgh, PA, USA

Andrew J. Strathern
Department of Anthropology
University of Pittsburgh
Pittsburgh, PA, USA

This book series addresses a timely and significant set of issues emergent from the study of Environmental [sometimes referred to as "natural"] disasters and the Series will also embrace works on Human-produced disasters (including both environmental and social impacts, e.g., migrations and displacements of humans). Topics such as climate change; social conflicts that result from forced re-settlement processes eventuating from environmental alterations, e.g., desertification shoreline loss, sinking islands, rising seas.

More information about this series at
http://www.palgrave.com/gp/series/15359

Jenny Bryant-Tokalau

Indigenous Pacific Approaches to Climate Change

Pacific Island Countries

palgrave
macmillan

Jenny Bryant-Tokalau
Te Tumu
University of Otago
Dunedin, New Zealand

Palgrave Studies in Disaster Anthropology
ISBN 978-3-319-78398-7   ISBN 978-3-319-78399-4   (eBook)
https://doi.org/10.1007/978-3-319-78399-4

Library of Congress Control Number: 2018940529

Cover design: Cover pattern © Harvey Loake

Printed on acid-free paper

This Palgrave Pivot imprint is published by the registered company Springer International
Publishing AG part of Springer Nature.
The registered company address is: Gewerbestrasse 11, 6330 Cham, Switzerland

To my son, Maretino Vuloaloa Bryant-Tokalau,
A true son of the Pacific

Series Editors' Preface

Knowledge, Experience and Policy: Environmental Issues in the Pacific Islands

All over the world environmental issues make themselves increasingly evident and pressing, posing multiple layers of problems of how to deal with them. Climate change appears to be a leading factor here, and there have also been notable changes in climate in prehistoric times, for example, with warming temperatures in the Pleistocene period some 10,000 years ago. In the Pacific, historical perturbations associated with the El Niño effect are well known. So, problems related to climate are by no means new. And problems do not simply result from climate change; they are always compounded by further human factors, such as overcrowding, poor location of residences, inadequate forms of construction and failure of governments or businesses to take notice of what local people with experience of managing climatic conditions over long periods of time are saying to them. The aphorism 'think global, act local' applies with considerable relevance here, because local means listening to what local people say and do.

We are very proud to be able to write this Preface in honour of the ground-breaking work of Professor Jenny Bryant-Tokalau, in which she consistently emphasizes the prime importance of incorporating indigenous knowledge into policies and actions designed to cope with developing environmental problems in the Pacific Islands. A mainstay of her knowledge is with the islands making up Fiji, were she has long-term field experience. Fiji recently (2016) suffered the massive affects of Cyclone Winston with widespread destruction of housing and urban as well as rural living spaces. In such circumstances, two considerations are paramount. One is to develop forms of replacement housing that will be better than temporary tents or shacks. The other is to re-establish patterns of self-sufficiency in terms of gardening and/or fishing so that people do not become dependent on aid hand-outs for their daily living. In both of these domains, incorporating local knowledge and skills is a crucial factor for recovery from environmental disasters. Such disasters do have the secondary effect of raising people's awareness of their ecological placement in the world, and this provides a context in which new patterns of activity, modelled on earlier ones, can come into being.

Quick and effective use of the resources already available is also important. In many Pacific Island places people know how to collect edible wild plants in case of garden failure. In some places, such as Vanuatu, certain types of indigenous houses are well adapted to withstanding the force of hurricanes or cyclones. In general, thatch-roofed wooden houses are less dangerous in a cyclone than metal-roofed dwellings. Thatch collapses, but corrugated iron blows around and can injure people. Knowledge is also connected to structures of community leadership and ties of kinship that can be mobilized in cases of disaster or emergency warnings of problems. In the Pacific, volcanic islands offer much better protection against flooding than do atolls, and people can use known pathways to reach higher land. Atolls are much more vulnerable because high sea waves can sometime engulf a whole island. Knowledge of higher spots that may escape such damage is again important, as also is seafaring skill when people find themselves in turbulent waters. Inland, knowledge of the location of potable spring water is a major asset, because of the need for rehydration while people search for temporary famine foods. Reading signs of weather trends in the appearance of rivers or the sea and in the movements of terrestrial animals or birds is also obviously crucial.

While urban life may obviously bring advantages in terms of services and food purchases on a daily basis, it also entails great dependence on external networks rather than internal self-sufficiency, so a major source of concern lies in the vulnerability of urban areas when the infrastructure of services collapses. In February 2016 we and a group of our students from the United States were informed by the Director of Emergency Services in the Cook Islands that in the main island of Rarotonga much traditional knowledge had been lost. By contrast, in the remote northern islands traditional environmental knowledge had, he explained, been maintained. In emergencies these islands could not easily be reached and therefore they had to rely on their own knowledge and methods of survival. Some of that knowledge could, however, be captured and shared between islands both within the Cook Islands and across other Pacific Island contexts. The University of the South Pacific, based in Fiji and with branches in the islands, could provide a home for such a databank of indigenous knowledge on the model of seed banks that house reserve collections of plant seeds to be drawn on if necessary.

Resilience in the face of environmental perturbations is likely to be strongly characteristic of kinship and community-based societies such as are widely found in the Pacific Islands. In Samoa, where a tsunami struck the southern coast in 2009, narratives of the response to this dangerous situation indicate clearly how a number of organizations worked together to promote escape from immediate danger and longer-term recovery. These organizations included churches, of all denominations, and the Red Cross. The strong network of kin groups, headed by chiefly leaders, provided the ongoing infrastructure and impetus assisting these organizations in their work. Indeed, local church congregations are themselves based in kin groups and their titled chiefs. Another factor fits in here. Church buildings tend to be strongly built, because having such buildings is a mark of community prestige and competition. These buildings can provide a refuge in cyclone conditions, as happened on Taveuni Island in Fiji during Cyclone Winston. Ordinary houses were extensively damaged or destroyed, but the church stood. The same pattern has happened in the Philippines where Catholic church buildings in cities are large and well built.

Climate change, throughout the Pacific and elsewhere, has increased the incidence of severe weather episodes, especially floods with high king tides and heavy rains. Cyclical patterns of major adverse events also affect countries around the Pacific Rim, such as Taiwan, where on 6 February 2018 onward severe earthquake shocks hit, especially in the city of Hualien. Events of this kind are not only traumatic in their immediate effects, but as with all disasters they are extremely costly in the longer run in terms of longer-term recovery.

In the course of her wide-ranging discussion, Professor Bryant-Tokalau touches on a long list of vital aspects of disaster recovery, mitigation and prevention. Food security is one topic, and she mentions burying bananas as a storage and protection device. The actual or potential role of the churches in ecological issues is another significant domain of issues. Long-term historical backgrounds are vital. Muriel Brookfield's detailed report on Cyclone Val's effects on Lakeba village in Fiji in 1975 carried numerous insights. The village had been free from cyclones for 27 years before the advent of Cyclone Val, so people were not ready for it. Lakeba was isolated, and communication with outside places was poor. After the disaster, many people just left the area, where 300 homes had been destroyed, and rebuilding them was not easy, even though Lakeba was the home village of Fiji's then President Ratu Sir Kamisese Mara. Overall responses were greatly inflected by patterns of community leadership. Ecological work to repair environments requires community coordination, consensus and sustained effort. On coastlines, sand dunes, grasses and especially mangrove swamps can provide maintenance against erosion and also valuable carbon sinks and living spaces for aquatic creatures. These resources have to be protected and enhanced. Here is where a classic topic in anthropological studies needs to be revived: kinship. Kinship, and its strategic extensions and transformations, is an overall vital key to social resilience in the face of environmental pressures, both in rural and in urban contexts.

Traditional ecological knowledge (TEK) is the other key to resilience, and Professor Bryant-Tokalau's book is explicitly written as a companion to another study for our series, by Dr. Lyn Carter, on Maori TEK (MEK) in Aotearoa New Zealand. These two studies will significantly help to focus more world understanding on the Pacific Islands and the vital knowledge that the study of them can contribute to disaster studies in general.

Pittsburgh, PA   Pamela J. Stewart (Strathern)
Pittsburgh, PA   Andrew J. Strathern

ABOUT THE SERIES EDITORS

Pamela J. Stewart (Strathern) and Andrew J. Strathern are a wife-and-husband research team who are based in the Department of Anthropology, University of Pittsburgh, USA, and co-direct the Cromie Burn Research Unit. They are frequently invited international lecturers and have worked with numbers of museums to assist them in documenting their collections. Stewart and Strathern have written over 50 books and over 250 articles, book chapters and essays on their research in the Pacific (mainly Papua New Guinea and the South-West Pacific region, e.g., Samoa and Fiji); Asia (mainly Taiwan, and also including Mainland China and Japan); and Europe (primarily Scotland, Ireland and the European Union countries in general); and also New Zealand and Australia. Their most recent co-authored books include *Witchcraft, Sorcery, Rumors, and Gossip* (2004); *Kinship in Action: Self and Group* (2011); *Peace-Making and the Imagination: Papua New Guinea Perspectives* (2011); *Ritual: Key Concepts* *in Religion* (2014); *Working in the Field: Anthropological Experiences Across the World* (Palgrave Macmillan, 2014) and *Breaking the Frames: Anthropological Conundrums* (Palgrave Macmillan, 2017). Their recent co-edited books include *The Ashgate Research Companion to Anthropology* (2015); *Exchange and Sacrifice* (2008) and *Religious and Ritual Change: Cosmologies and Histories* (2009 and the Updated and Revised Chinese version: Taipei, Taiwan: Linking Publishing, 2010). Stewart and Strathern's current research includes the topics of cosmological landscapes; ritual studies; political peace-making; comparative anthropological studies of disasters and climatic change; language, culture and cognitive science; and Scottish and Irish studies. For many years they served as Associate Editor and General Editor (respectively) for the *Association for Social Anthropology in Oceania* book series and they are Co-Series Editors for the *Anthropology and Cultural History in Asia and the Indo-Pacific* book series. They also are currently co-editing five book series: *Ritual Studies; Medical Anthropology; European Anthropology; Disaster Anthropology* and *Anthropology and Cultural History in Asia and the Indo-Pacific* and they are the long-standing co-editors of the *Journal of Ritual Studies* [Facebook: https://www.facebook.com/ritualstudies]. Their webpages, listing publications and other scholarly activities, are: http://www.pitt.edu/~strather/ and http://www.StewartStrathern.pitt.edu/.

PREFACE

Although the Islands of the Pacific are constantly portrayed as victims in the face of climate change, and indeed are facing many major impacts with increasingly strong storms and rising sea levels, they also have many ways of facing up to the changes. Pacific Islanders have much to teach other nations about resilience and coping and are determined to use their knowledge to maintain ancient traditional practices.

This book, by examining the many ways that people respond and adapt, shows that while there is much to do in the face of what is now inevitable change, that learning from the Pacific rather than imposing all technological fixes from outside, can be beneficial to all communities.

Dunedin, New Zealand   Jenny Bryant-Tokalau

Acknowledgements

This book has been some time in the making. Lyn Carter and I had the idea for a combined work during 2015 but on completion of an initial draft, along with growing global knowledge and acceptance of climate change, as well as different approaches in the wider Pacific, it was decided instead to produce a pair of companion books. These 'companions' should be read together as they demonstrate the different approaches taken by the island Pacific, and one of their neighbours, Aotearoa New Zealand.

During the time of writing I have been employed in Te Tumu, the Department of Maori, Pacific and Indigenous Studies at the University of Otago. In 2016 during an abbreviated sabbatical I was able to work on some aspects of the book, but mostly it has been written in 'spare' hours. The work itself has drawn upon some of my earlier career with the University of the South Pacific in Suva where I taught Geography, and with UNDP Pacific where, as a Global Environment Facility staffer, I was involved closely with early projects on climate change adaptation and mitigation. Many of my former students and colleagues from those days have been involved with my examination of and discussion of issues surrounding traditional adaptations, and have shared in my disappointment that so much of donor assistance has failed to take into account the traditions, knowledge and practices of Pacific Islanders who have for centuries known how to adapt and survive. Despite exposure of much of the wastage and repetition in donor funding, Pacific people's voices are not always acknowledged.

Several individuals have however kept those voices in the forefront. In Fiji, Joeli Veitayaki, and our late and sadly missed colleague Lionel Gibson and I had many debates and discussions over what happens in the 'aid game'. Aliti Vunisea, good friend and colleague, has always been my guide and support; from Kiribati, Naomi Biribo, former student and colleague, possibly unknowingly tipped me in this direction when she showed me a film on coral sand mining; from Solomon Islands, Irene Hundleby with her energy and huge interest has helped in many ways, including providing some of the photographs; Geographers Tim Bayliss-Smith and William C. Clarke were always there to read and suggest useful background material, and encouraged me to go back to the early work of Muriel Brookfield on cyclone preparation. In New Zealand and the Pacific, recent works of Bedford, Campbell, Rakova, Tabe and others, as well as suggestions from anonymous peer reviewers have been valuable. Between Fiji and New Zealand, my late husband, Filipo, and our son Maretino have always been hugely supportive of my work and travel habits, providing constant reminders to 'take it easy'. Vina'a va'a levu saroga.

CONTENTS

Acronyms

ADB	Asian Development Bank
A/NZ	Aotearoa New Zealand
AOSIS	Alliance of Small Island States
ASPEI	Association of South Pacific Environmental Institutions
CCS	Carbon Capture and Storage
CITES	Convention on International Trade in Endangered Species of Wild Fauna and Flora
DRR	Disaster Risk Reduction
EEZ	Exclusive Economic Zone
EMSEC	Emergency Management and Security Committee (Fiji)
ESCAP	The United Nations Economic and Social Commission for Asia and the Pacific
ETS	Emissions Trading Scheme
FRIEND	Foundation for Rural Integrated Enterprises and Development (Fiji)
GEF	Global Environment Facility
GHG	Greenhouse Gases
ILK	Indigenous Knowledge Systems
IPCC	Intergovernmental Panel on Climate Change
MAB	Man and the Biosphere Programme (UNESCO)
MEK	Maori Environmental Knowledge
MfE	Ministry for Environment
NEMS	National Environmental Management Strategies
NGOs	Non-government Organizations
NIMBY	Not In My Back Yard

NIWA	National Institute of Water & Atmospheric Research
NZETS	New Zealand Emissions Trading Scheme
PICs	Pacific Island Countries
PIDF	Pacific Island Development Forum
PIF	Pacific Islands Forum
PIPP	Pacific Institute of Public Policy
PNG	Papua New Guinea
PSIDS	Pacific Small Island Developing States
REDD	Reducing emissions from deforestation and forest degradation
SBTF	Sovi Basin Trust Fund
SCBD	Secretariat of the Convention on Biological Diversity
SDGs	Sustainable Development Goals
SIDS	Small Island Developing States
SME	Small to Medium Enterprises
SOPAC	Pacific Islands Applied Geoscience Commission
SPC	Secretariat of the Pacific Community
SPREP	Secretariat of the Pacific Regional Environment Programme
TEK	Traditional Environmental Knowledge
UNCED	United Nations Conference on Environment and Development
UNDP	United Nations Development Programme
UNEP	United Nations Environment Programme
UNESCO	United Nations Educational, Scientific and Cultural Organisation
UNFCCC	United Nations Framework Convention on Climate Change
USP	University of the South Pacific

LIST OF FIGURES

Introduction

Abstract In this introduction Pacific and global concerns around climate change and its impacts are outlined. The need for a deeper understanding of the interrelatedness of local expertise, customary knowledge and practice, and traditional adaptive responses is emphasized with a focus on sea-level rise and flooding. Chapter 1 demonstrates why it is necessary to have a pair of books, one looking specifically at the island Pacific and the other on Aotearoa New Zealand, in order to demonstrate that there are important lessons to be learned from Pacific Islanders. These lessons on how it may be possible to adapt, using traditional ecological knowledge, to the impacts of climate change apply not only to Aotearoa New Zealand but also to the wider world.

Keywords TEK (traditional ecological knowledge) • PICs (Pacific Island countries) • Adaptation • Diversity • Responses • Resilience

Most people with an interest in climate change will be familiar with the poem 'Dear Matafele Peinem' by Kathy Jetnil-Kijiner, from the Marshall Islands, who passionately narrated it in her address to the Opening Ceremony of the UN Secretary General's Climate Summit on 23 September 2014. Many were moved to tears imagining (and some had indeed experienced) the personal impacts of climate change on the people of the Pacific and across the globe. Kathy's line *no one's gonna become a*

© The Author(s) 2018

J. Bryant-Tokalau, *Indigenous Pacific Approaches to Climate Change*, Palgrave Studies in Disaster Anthropology, https://doi.org/10.1007/978-3-319-78399-4_1

climate change refugee[1] became a refrain for Pacific climate change movements. Globally people were deeply moved by the determination and the refusal to succumb.

In this book focusing on the island Pacific, knowledge and general acceptance of how climate change issues are affecting the Pacific region are assumed. Almost everywhere it is understood that globally there is an urgent need to limit human-induced temperature rise, slow rising sea levels and increasingly intense flooding, storms and droughts. There are many areas of grave concern. Fresh water accessibility and growing threats of disease are now understood to be immediate security threats to the sustainability of Pacific populations and nowhere is immune. Crop failures and thus food security, limited water supplies and rising temperatures are all now being felt. Significantly however, the peoples of the Pacific Islands are not only aware of changing climate and its impacts, but are constantly working (at all levels) to face the challenges before them. This is happening not only at home but also on the global stage.

Throughout the world, from the Middle East to Asia, the Americas and African and European continents, environmental hazards, major floods and other climate-related disasters are growing in intensity. Globally, they also disproportionally affect the poor and marginal. The island Pacific nations are not alone in this, but what is relevant and may be learned from this book is recognition that traditional and community responses and understandings of climate and disasters are still widely understood and practiced. People across the Pacific are now taking extreme events more seriously, responding through education, sharing and historical knowledge. Recognition of such knowledge in other parts of the world may well be another approach to dealing with disasters. Community cooperation and participation in decision-making as well as recognition and promotion of forms of community resilience are found everywhere and need to be incorporated in disaster response at both institutional and community levels. This book demonstrates that by understanding the interrelatedness of local expertise, customary resource management, knowledge and practice, as well as the roles of leaders and institutions, local 'knowledge-practice-belief systems' can be used to inform adaptation to disasters wherever they occur.

Very different to the situation in larger countries, such as Aotearoa New Zealand, in the island Pacific (the Pacific Island countries [PICs]) there is more unity about climate change. Populations do not need to be convinced of what is happening and there are very few sceptics and naysayers. As demonstrated in this book, this is very evident in the respect for and continuing use of traditional ecological knowledge (TEK) in countries that are largely independent and governed by indigenous communities.

This book on Pacific approaches to climate change provides lessons on what Pacific countries can teach other nations, especially Aotearoa New Zealand. It will show that, far from what is often portrayed in the media, islanders and their countries are not always as vulnerable as they may appear, and had, in the past, the ability to survive in the face of environmental changes without a large amount of assistance from donors. These qualities are once more being given prominence in adapting to the growing impacts of change. Perceptions of vulnerability can increase feelings of disempowerment, and yet these are qualities not widespread in the Pacific at present. As Kathy Jetnil-Kijiner so eloquently said, Pacific islanders are not only fighting back utilizing traditional and contemporary resilience to inundation, flooding and more intense storms, but also responding as they always have to threats of loss of territory and the impacts of devastating winds and flooding by using their local knowledge and innovation as well as modern technology. There is very little possibility that the majority of PICs can afford many of the technologies required to protect the coasts, for example, so they must utilize donor assistance. That assistance will be far more effective if it is locally appropriate and draws upon existing knowledge.[2]

Climate change is a significant environmental and human security issue. As Pacific countries already demonstrate that they are skilled at adaptation, then the incorporation of both traditional and new practices would enhance people's ability to adapt. Yet the pace of climate change is now such that even traditional resilience and adaptation utilized by communities and governments are being severely tested. As has been seen with recent storms across the globe, some highly technical fixes such as engineering solutions of seawalls and complex artificial islands can risk failure if not carefully planned.

It is important to listen to local communities and to respect the knowledge that exists. It is argued here that Pacific islanders have been adapting to and mitigating against environmental change for much longer than is currently understood. Travel over vast distances to reach their now homelands, regular movements back and forth between islands, dealing with disasters, sudden environmental change and also incremental change over decades, all beyond the control of small communities, do eventually lead to adaptation.

It is not easy however, and some of the 'super storms' such as Cyclone Winston in Fiji in 2016 have almost overwhelmed the ability of people to cope. While cyclones are an anticipated part of seasonal life in the Pacific, storms of growing intensity in recent years have meant that even traditional ways of coping are being challenged, but the models of these methods such as adaptation through relocating, or the construction of artificial islands, and building upon traditional relationships, can provide good lessons for coping with climate and indeed other forms of environmental change. Pacific Island countries are thus continuing to contribute to global research and policy to 'turn the tide'. It may be too late for some low-lying islands, but active engagement in seeking actions aimed at 'safeguarding biodiversity and ecosystems; ensuring food, water and energy security; and supporting future socio-economic development by becoming climate resilient' is ongoing.

This is a companion book to Lyn Carter's *Indigenous Approaches to Climate Change: Aotearoa New Zealand* that aims to demonstrate how Aotearoa New Zealand can benefit from the many Pacific adaptation strategies already in place. The focus in both books is on how TEK informs adaptation strategies and practices and can be shared with Aotearoa New Zealand, and other countries, by using TEK frameworks in dealing with climate change.

The many existing and historic adaptation examples from the island Pacific demonstrate that such frameworks are not only remembered and daily replicated, but also still have great significance in dealing with the current global focus on climatic change and all its implications. The capacity of the world to adapt may prove to be great, yet will also be a challenge. The greatest impact is likely to be for indigenous communities, the growing poor and impoverished, and for groups whose lives are closely linked to their local environments. Indigenous peoples who are part of their local ecosystems will experience substantial challenges to their lives, involving not only loss of habitat but also declining food security. In the PICs where people accept the fact that the climate is changing, and are attempting to find ways of dealing with such changes, there are many lessons to be shared including with larger Pacific neighbours such as Aotearoa New Zealand (Fig. 1.1).

In Lyn Carter's companion book, she provides a detailed account of TEK and practices throughout the Pacific that have been developed over time to ward off historic environmental challenges and disasters. She

Fig. 1.1 Map of Pacific including Aotearoa New Zealand. Source: https://en.wikipedia.org/wiki/File:South-pacific-map.jpg

notes how, from the times of early voyages across our Oceanic highways, PICs have been adapting to and mitigating against many forms of environmental change and stresses the influence of culture and cultural landscapes that have endured across time and can provide a basis for communities facing the challenges of climate change. The understanding that Pacific peoples have of the integration of their human, physical and spiritual worlds informs their knowledge frameworks, even when increasing urbanization, higher technology and industrial agriculture (in some countries) are not always discussed or actively considered. For many in Pacific countries TEK simply *is*; it is almost an unconscious part of everyday life, increasingly remembered as environmental challenges become great. As Carter says, 'the knowledge frameworks and processes that are being utilized are TEK based and Pacific worldviews have been shaped over time through people's interaction with their environment'.

In this book, located entirely within the island Pacific, TEK need not be described again, but is understood to be what informs 'locally referenced, experiential knowledge, practices and solutions ... [to] demonstrate the adaptive capability for indigenous peoples to ensure long-term sustainable use and habitation within their cultural landscapes' (Carter).

In these books on *Indigenous Approaches to Climate Change*, we have elected to focus on two areas of climate change induced challenges, sea-level rise and flooding, highlighting examples of adaptation measures currently in place or planned in the Pacific, including New Zealand. Because of the focus on these areas in the 2014 Intergovernmental Panel on Climate Change (IPCC) report, we have chosen not to specifically discuss energy projects designed to reduce the reliance on fossil fuels. We do acknowledge however the importance of this adaptation strategy for the wider Pacific region where there is a great deal happening around energy alternatives. There is a vast amount of literature and research around climate change and fossil fuel reduction, whereas positive stories about adaptation measures that focus on the Pacific challenges from sea-level rise and flooding are less prolific. Taking the lead from the IPCC, the case studies we have chosen to include here will be detailed studies on existing adaptation measures that contribute to minimizing the effects of sea-level rise and flooding. We shall not be discussing scientific evidence for climate change but have chosen to accept it.[3]

THE PACIFIC BOOK

In this book on the Pacific Islands, case studies from Solomon Islands, Vanuatu, Fiji and Kiribati, along with the work and approaches of several Pacific regional institutions and agreements are presented as examples of how countries are responding to and learning about climate change, not only from international scientists, donors, individuals and their knowledge, but also from generations of traditional adaptation, journeys, practices and canny global, regional and national politics. The deliberately detailed case studies given here are often very country or locality specific. They have been carefully chosen to illustrate a range of indigenous responses to a changing climate and to provide lessons for other nations who may also wish to listen to their own indigenous voices that have long adapted to the inevitability of a changing climate.

In this pair of books, the term PICs[4] refers to those that are on the whole governed by the indigenous peoples. The countries that make up the island Pacific are extremely diverse: culturally, linguistically, historically and politically. Within each country there is also much diversity not only in culture and language but also in how people respond to challenges and pressures, modern development and change. It follows then that responses and reactions to the impacts from climate change are also very diverse and develop from the individual country's cultural, political, economic and environmental landscape. Importantly however, the countries are also united in a common Pacific stance to keep 'climate change very high up on global political agenda'. In Aotearoa New Zealand this is not the case even though indigenous knowledge is incorporated into overall government policy and action in ways that differ to the diverse experiences in the PICs. This Pacific-focused book is intended to provide examples of good lessons that many in Aotearoa New Zealand are unaware of and may be interested to emulate, or at least recognize and support.

In Chap. 2, 'Pacific Responses to and Knowledge of Climate Change: Institutions and Tradition', I examine academic and institutional history and approaches to climate change in the Pacific Islands. Leading up to the Small Islands Developing States (SIDS) conference held in Samoa in 2014 where the emphasis was very firmly on the challenges of climate change to low-lying islands and coastal areas, I trace a history of formal environmental management since the 1950s with the establishment of regional organizations, the impact of the Pacific's difficult nuclear history on these organizations and the shift to other concerns such as biodiversity, waste and climate. Global institutional developments within the United Nations (UN) and other multilateral organizations have had an enormous influence on the way that PICs responded to their many environmental concerns, and indeed caused much stress in terms of dealing with institutional demands. As demonstrated here however, the PICs themselves have also had some influence on global practice, particularly through alliances such as Small Island States and the United Nations Framework Convention on Climate Change (UNFCCC). Pacific diplomacy changes over time of course according to the needs and alliances of the diverse group of countries. Some regional institutions have been developed around particular issues, and the power and focus of those institutions has increased and diminished over time and in relation to global and regional politics. Fry and Tarte eds. (2016) in *The New Pacific Diplomacy* discuss the engagement of nations at regional and global levels, and the 'paradigm shift' that is occurring.[5]

Institutions do not only involve large, international bodies. Chapter 2 also sets the scene for later chapters where formal religion, spirituality and the fundamental belief systems of the many, complex Pacific societies are shown to be intrinsic to the multiple ways that communities, individuals and governments respond to the challenges of climate change. Many of these have evolved over centuries and are often more appropriate than the current 'technical fix' responses to inundation, droughts and major storms. In Chap. 3, 'Adaptation to Climate Change in the Pacific Islands: Theory, Dreams, Practice and Reality', examples of contemporary donor and business responses to inundation such as the construction of artificial islands are illustrated with a range of developments, including the 'fantastic' engineered floating islands approach, to recent coastal tourist developments and the latest plans for floating cities in French Polynesia. A key point in this section is that Pacific islanders have always constructed artificial islands, as a response both to a shortage of land living space and for cultural, relational and environmental reasons. The long history of such developments and traditional adaptations that communities have always and still make to changing environments is illustrated with cases from Solomon Islands, Vanuatu and areas of Micronesia. Inevitably, some communities have had to relocate from their homelands, and it may be that this will become more usual as solutions involving construction of seawalls, for example, are no longer adequate. Issues surrounding relocation are examined in the second part of Chap. 3, which looks specifically at the well-known and widely reported situation of the islands of Kiribati. Some of the representations of Kiribati and climate change impacts are challenged in this chapter, especially widely presented beliefs and portrayals that what is happening now is new and will lead ultimately to the disappearance of an entire country of 'victims'. Using evidence from oceanic geomorphologists as well as local i-Kiribati scientists and other Pacific academics, alternative scenarios for the country are outlined. The option of relocation, particularly the purchase of land in Fiji is discussed in some depth with several scenarios highlighted. The land purchase is variously presented as having been wise as it will provide food security, or ill-advised, unsuitable for agriculture and settlement, and perhaps most brutally, likely to cause loss of security for another of Fiji's migrant peoples, the Solomon Islanders, originally brought to Fiji as 'black birded' labourers. It is also suggested that what may make more sense for the peoples of Kiribati is internal relocation. Whatever the result, and the resilience of people to an obviously uncertain future, it is clear that the people themselves need to have more say over what happens next.

A theme throughout this book is how populations respond to extreme events now and in earlier times. Migration and resettlement to other islands and lands is not always considered or even possible, and local responses to flooding and increased hurricane activity are detailed in Chap. 4, 'Handling Weather Disasters: Resilience and Adaptive Capacity of Communities', which examines the adaptations that communities make in their changing environments, especially when faced with extreme weather disasters. The recent case of Cyclone Winston in Fiji contrasted with an earlier situation following Cyclone Val highlights how traditional adaptations can mean long-term survival for communities, especially when no assistance arrives for days or even weeks. Taking Fiji as a particular case, the development of new, local organizations, working with villagers and local communities, demonstrates how resilience can be increased in times of uncertainty and disasters of almost unimaginable proportions.

In Chap. 5, 'Indigenous Knowledge Systems and Urbanization: Relocation, Planning and Modern Disasters', how people prepare for, adapt to and demonstrate their resilience in urban settings are examined. To this point this book has focused largely on rural, isolated and island communities, but the Pacific is rapidly urbanizing with as many as 50 per cent of populations living in towns and cities. It is accepted that there is a long history in the Pacific Islands of indigenous recognition of changing weather patterns and making preparations in advance of extreme events, but what is less well understood is that in societies where not all people are indigenous, and where a growing proportion are clearly urban, and frequently away from traditional networks, forms of adaptation and resilience continue. As this chapter demonstrates, urbanization does not necessarily mean the casting off or loss of knowledge of how to cope. Ikeda's mantra of robust systems, resourcefulness, rapidity and redundancy are referred to here as they are all to be found in Pacific societies, including in urban areas. As noted in the works of George Carter, for example, people do not always rely on government institutions to survive, and nor can they wait for assistance in the face of immediate danger. Pacific voices are found at all levels, not only in formal, planning settings. Recognition of indigenous knowledge systems is highlighted in this chapter as fundamental to disaster management, including in urban areas where people (no matter what their 'indigeneity') are developing new networks, indigenizing the towns and working as communities. It is argued here that urban dwellers and the poor are no less vulnerable than rural and island dwellers and that the patterns of resilience need to be utilized to inform adaptation to climate change disasters.

Finally, in Chap. 6, the concluding chapter, 'What Can Pacific Island Countries Teach Aotearoa About Climate Change?', it is opportune to reflect on dynamic, traditional and contemporary approaches to a changing world where it is now widely understood and experienced that the climate really is changing. This section draws the preceding chapters together and, as with Carter's companion book, intends to open the way for further discussions and debate on Pacific adaptation strategies and policies and how these may provide lessons for other nations such as Aotearoa New Zealand. Part of the discussion reviews examples of 'thinking outside the square' with more emphasis on recognizing that Pacific economies are 'blue' as much as they are green, and support of ocean environments is as important as those based on land, especially as they cannot be separated. The case of the Sovi Basin as an example of the thinking around the developing science for blue carbon and its potential for inclusion into mitigation strategies (such as Emission Trading schemes) across the Pacific demonstrate that, like all ongoing environmental challenges in the Pacific, climate change mitigation and adaptation are forever evolving. The adaptation of old technologies, such as sailing ships, to modern circumstances as a counter to the high dependence of economies so dependent upon fossil fuels demonstrate how well the TEK philosophy of the past can inform the present and develop intergenerational practices to ensure sustainability and growth. It also demonstrates how Pacific peoples understand 'in profound and elegant ways that we are all related'[6] and that the land-sea interface is an ever changing and challenging space in our lives. In emphasizing new forms of adaptation made possible through modern technologies, the concluding chapter represents what the many diverse nations and groups of Pacific islanders are doing for themselves and what lessons may be learned from the diversity of approaches. There are many lessons in these examples for the wider Pacific region, especially Aotearoa New Zealand.

NOTES

1. https://citizenactionmonitor.wordpress.com/. The poem was first read at the Opening Ceremony of the UN Secretary-General's Climate Summit, September 2014.
2. It is not the role of this book to examine in depth the appropriateness or otherwise of donor assistance in the Pacific. There have been many analyses in the Pacific, including works by Bryant-Tokalau (2008) and the Pacific Institute of Public Policy (PIPP).
3. There are many forms of environmental change and it is well recognized that not all dramatic change is wholly climate related. Populations everywhere do sometimes confuse causes of such change but most people are also aware of the damage that they can precipitate through poor land management, rapid deforestation and so on. For the purposes of this book, impacts of changing climates on the livelihoods and futures of Pacific peoples are the focus.
4. PICs covered in this book include the countries that make up the Pacific Island Forum: a political grouping of 16 independent and self-governing states. Members include Cook Islands, Federated States of Micronesia, Fiji, Kiribati, Nauru, Niue, Palau, Papua New Guinea, Republic of Marshall Islands, Samoa, Solomon Islands, Tonga, Tuvalu and Vanuatu, as well as Australia and New Zealand. The book does not cover the Pacific Rim countries.
5. Fry and Tarte (2016) in their edited volume provide an excellent analysis of negotiations at all levels around climate change (as well as other areas).
6. Cajete (2000), p. 178.

CHAPTER 2

Pacific Responses to and Knowledge of Climate Change

Abstract In this chapter, a brief history of environmental management within the Pacific Islands is presented with strong emphasis on the linkages between good environmental management as a key to economic and human development. Readers are reminded not only of the ancient and long journeys carried out by Pacific peoples to reach where they live today, but also of recent events that have galvanized people into seeking more prominent voices on key issues such as nuclear testing and the need to protect themselves against newer environmental challenges such as deforestation and mining. Voices of locals, communities, academics, and institutional, spiritual, regional bodies and global approaches are presented here.

Keywords Regional • Donors • Spirituality • Institutions • SPREP • Community

A Brief History of Pacific Environmental Management

PICs have been variously called 'small', 'vulnerable', 'isolated', 'geographically challenged' and 'poor'.[1] While some of these labels may be true for some islands (but certainly not all), and some have resources beyond the imagination of Aotearoa New Zealand, external perceptions of the Pacific Islands can belie their capacity to manage themselves, sometimes manipulate

© The Author(s) 2018
J. Bryant-Tokalau, *Indigenous Pacific Approaches to Climate Change*, Palgrave Studies in Disaster Anthropology,
https://doi.org/10.1007/978-3-319-78399-4_2

donor communities, and in recent times, have some influence on the global stage. Although influence may be overstating the case in these increasingly globalized times, there is no doubt that Pacific nations are not completely helpless and can provide lessons, particularly lessons in using TEK, to other nations.

Pacific countries have long been aware of a range of approaches to good environmental management as key to economic and human development. It is demonstrated in this chapter that Pacific Islanders have been adapting to, and mitigating against, environmental change for much longer than is often currently understood. Travel over vast distances to reach their now homelands, regular movements back and forth between islands, dealing with disasters, sudden environmental change and also incremental change over decades, all beyond the control of small communities, eventually lead to adaptation.

Initially the regional meetings of Pacific countries largely came about through the efforts of the early colonial powers to form groupings of nations to deal with regional issues, such as health. Countries have been meeting annually since 1950 to discuss common development issues.[2] Early concerns focused around atmospheric nuclear testing on Bikini and Eniwetok in the Marshall Islands, and then later French nuclear testing in French Polynesia. Nuclear 'testing' brought nations together in opposition to the tests and to the enormous social and health impacts on the peoples of the Pacific. The history of nuclear 'testing' is well known throughout the island Pacific as well as Aotearoa New Zealand, and there were many individuals and communities who actively opposed the tests. But at the governmental level, much of the opposition was informal because of the exclusion of 'political' issues from debate in organizations such as the South Pacific Commission.[3] It wasn't until 1970 that nuclear testing was formally on the agenda.

By the 1970s Pacific countries were very aware of the need for regional cooperation over a range of environmental issues. Several significant events occurred in this decade.[4,5] In 1971 a regional symposium on the Conservation of Nature was convened by the South Pacific Commission and the International Union for the Conservation of Nature and Natural Resources (IUCN). It could be said that this initial symposium led to the idea for a regional organization managed by PICs and, in 1974 the idea for a South Pacific regional environmental management programme was conceived. As an initial contribution to an understanding of how local environments were faring, each nation was asked to prepare a country report on the state of their environment. These reports, all published around 1973/1975, became the basis of an Action Plan and the Convention for the Protection of Natural Resources and Protocols on pollution and dumping in the Pacific.

For several decades, countries have developed institutionalized environmental planning and management with supporting legislation and regulations and all are recipients of aid and technical assistance from bilateral and multilateral donors as well as global non-government organizations (NGOs). From the 1980s each country has formulated a wide range of reports, for example, the National Environmental Management Strategies (NEMS)[6] and Capacity 21 projects.[7]

Pacific countries[8] are thus very well versed in writing reports and enacting legislation, dealing with aid donors and making plans in order to achieve, or at least strive for, sustainable, people-centred development. But this requires much more than the preparation of reports. It should also involve the commitment of traditional Pacific Island community organizations, such as customary land-owning groups, and although this is widely done, it is not always the case. The role of the churches and other traditional belief systems are also under-recognized in formal responses to climate change. An understanding of spirituality, indigenous practices as well as customary rights over land and marine resources and local environmental knowledge are more likely to ensure success of any legislation and externally generated plans. The Pacific countries provide many examples, as will be demonstrated here.

Today, 25 years after the Earth Summit in Rio de Janeiro, where climate change was barely on the agenda, and where the UN and thousands of people from all walks of life gathered to encourage governments and communities to rethink economic development,[9] it is useful to understand how Pacific countries are dealing with climate change and what they have learned from the decades of aid, consultants and legislation. Most crucially, we need to understand how widespread is the understanding of environmental change throughout Pacific communities. A Pacific response is not always one of unity. The region is very diverse with complex origins, historical events, networks and beliefs.

The nations of the Pacific extend across 29,500,000 square kilometres of ocean with a total land area of 550,044 square kilometres. Eighty-four per cent of that land area is in Papua New Guinea (PNG). Populations range from as small as 1500 people (Niue) to PNG at close to 8 million. Land areas also vary enormously from small, low-lying coral atolls to raised limestone platforms and high volcanic islands. The level of biodiversity is high. On the volcanic, high islands, such as Fiji, Papua New Guinea and Solomon Islands, there are vast forestry and mineral resources but the potential for exploitation is great. Resource rich nations are always in danger of downstream impacts of logging and mining such as the choking of reefs, loss of coastal fisheries and inequitable distribution of the financial rewards from logging, community conflicts, corruption and dramatic social, economic and political change, caused largely by the potential value of the forests.

Similar difficulties apply to mining, particularly in PNG, but also in Fiji, New Caledonia and Solomon Islands. Fisheries resources across the region are under threat from over-fishing, particularly from 'foreign' fishing fleets that take advantage of the vast Pacific Ocean that is largely impossible to monitor by either sea or air.[10]

It is not only human-induced disaster and disruption that affects the nations of the Pacific. Natural damage from cyclones, earthquakes, volcanoes, floods, forest fires from lightning strikes, droughts and frosts can also have major impacts on resource availability, but is increasingly understood that such 'natural' disasters are intensified by human actions. Given the many environmental challenges that face Pacific nations, the now well-recognized threat of climate change and sea-level rise will impact on the very future of these countries and populations.

In late August 2014 around 3000 delegates attended the Third International Conference on Small Island Developing States (SIDS) in Samoa. This was a highly significant meeting, following on from those in Barbados and Mauritius.[11] Most important was not only that it was held in the Pacific Islands, but that it also emphasized challenges wrought by climate change on those small and low-lying islands and coasts. The event was addressed by United Nations Secretary General Ban Ki-Moon who stressed that this meeting was a 'once-in-a-decade opportunity to promote sustainable development in countries such as Samoa, which are often described as on the front line of climate change'.[12]

I know your country is facing a lot of difficulties. First of all, by climate change, rising sea tides. Low-lying island nations, some of which are little more than one meter above sea level, are regarded as some of the most vulnerable to rising seas blamed on man-made climate change. Some small states in the Pacific such as Kiribati have already begun examining options for their people if climate change forces them from their homeland.

Other delegates, such as the World Bank special envoy for climate change Rachel Kyte, noted that island nations did not create the problem and did not have the resources to deal with it, so the rest of the world was obliged to help. 'We have a responsibility towards these nations because we've pumped enough poison into the atmosphere over the last decades to imperil the livelihoods of many of the people in many of the atolls and islands of these nations.'[13]

But the responses, both from donors and from the countries themselves, are necessarily diverse. Academic analysis, institutional concerns, churches, the governments and the populations may largely agree on the threats and potential impacts of climate change, but they also have individual responses and will form coalitions to advance their interests.[14] Goulding (2016) examines regional institutions and shows how complex it is to have a regional voice. There are other participants as well, and the role of research interests, funding, spirituality and formal religion, as well as the history of people and communities should not be overlooked.

ACADEMIC APPROACHES TO CLIMATE CHANGE

In 1991 Patrick Nunn, then a Geographer at the University of the South Pacific in Fiji, published a small monograph on *Human and Nonhuman Impacts on Pacific Island Environments*. The focus (and subtitle) of the monograph was '*feeling the hand of god*' alluding to the role of god, and/ or an unknown deity in determining the fate of land, environment and climate. Nunn, like most geographers throughout history, recognized the impacts of humans on environments, but reminded readers that no matter what humans do, environmental changes still occur. As a geomorphologist, Nunn, while recognizing human impact associated with modern developments, was then largely concerned with emphasizing the environmental and climatic changes that have occurred in the Pacific since the Quaternary[15] period, changes not dissimilar from those that have occurred globally.[16]

Nunn's aim was to point out that long-term environmental change has been significant, and that 'these changes did not abruptly cease once humans settled the islands'.[17] He was concerned, in essence, that the physical geologic record (as well as pollen and other climatic changes) would not be forgotten in the then relatively new debates on Pacific climate change, its impacts and the emphasis on the role of humans. Nunn's ideas supported views that the world recognizes and acts upon the role of humans in environmental (and particularly climate) change, and he played a valuable part in reminding the reading public that the science of climate change, and people's understanding of that change, is far more complex than is often portrayed. Nunn was not alone, of course. Many have challenged the fundamental causes of climate change, climate science and the notion of 'who is responsible'. Now, more than 25 years after the publication of Nunn's monograph, the debate has reached the point where scientific projections of 2 degrees Celsius rise in global temperatures, sea-level rise with particular impacts on small, low-lying islands and coastal areas, along with an increasing intensity of storms, has the world scrambling for solutions. Pacific Islanders are in all camps. Many understand and are part of the search for scientific knowledge, they are aware of the depth of the geological, archaeological and climatic records, but also believe in *the hand of god*. '*Isa*, what will be, will be' is a common cry. Blame for the traumatic changes facing the globe is still being apportioned, but largely ways of facing and dealing with these changes (adaptation and mitigation in UN parlance) are not just being sought, they are also underway.

Although he made no reference to them, Nunn may well have been responding to earlier reports by Secretariat of the Pacific Regional Environmental Programme (SPREP) and the Association of South Pacific Environmental Institutions (ASPEI)[18,19] and in particular a small document by Hulm, 'Climate of Crisis' which spoke of global warming threatening the survival of Pacific Islands and their cultures. The focus was on the impacts of carbon dioxide concentrations in the earth's atmosphere and its likely impacts. Hulm (and the substantial report from which the booklet was drawn[20]) spoke of resettlement, loss of resources, fragile ecosystems and the potential change in rights to marine resources, small islands and Exclusive Economic Zones (EEZs). Working on a scenario of 2 degrees rise in temperature by the year 2100, and a sea-level rise by 1 metre by 2050 the reports outlined scenarios and gave advice on how small Pacific Island groupings should respond. Flooding, disease, increased humidity, droughts, loss of arable land, marine and land-based species, an increase in invasive species and a loss of forests were all predicted. The resettlement of groups of people, such as from the Carteret Islands (who in fact had agreed to resettlement as early as the 1960s due to the loss of gardening land from heavy seas) were all discussed in some depth. Nunn's arguments were that such environmental change is normal to the earth's history and has been happening for millennia.

ASPEI and SPREP were pointing to human-induced change, and no matter what terminology was then used, whether it was 'global warming', climate change or simply environmental change, the world's larger populations have been slower than the Pacific in recognizing the implications of what has now become rapid, and not immediately reversible, climate change, with the 2 degrees temperature increase imminent, more intense weather patterns now upon us, and the danger of geographical shifts in the patterns of invasive species and disease occurring now. Geomorphologists and archaeologists will of course remind us, as did Nunn, that such change is part of the human and pre-human condition to life on earth. What is new(er) is how humans adapt and why they have to respond more rapidly, within a period of one or two generations and not thousands of years. No matter who or what is at fault, residents of small islands and low-lying coastal areas need to adapt, and mitigate, now and not later.

Since these early works on Pacific climate change there has been a significant production of academic research and writing. Geographers and anthropologists including Barnett and Waters, Connell, Hviding and Crook have published widely on the growing 'vulnerability' of Pacific Islanders,[21] to the impacts of climate change, especially rising sea levels, impacts on agriculture and loss of land leading to out-migration. Others have focused on the role of Pacific countries at the international level where the 'Pacific voice' has become one of apparent unity, but where individual country and cultural group responses vary. Representation of the adaptive capacity of Pacific Islanders is still sometimes presented as entirely negative, such as in the pamphlet by Bell et al. on Climate Change and Pacific Island Food Systems,[22] but largely in more recent years, as presented in the case studies, Pacific voices such as those of G. Carter[23] have become more nuanced with indigenous academics challenging and extending representations of vulnerability and victimhood. Their writings present the academic audience with many voices, ranging from those of atoll groupings in the global world of climate change diplomacy to digital coalitions with Pacific negotiators using the electronic world to great advantage producing a successful form of e-diplomacy (see Carter 2015: 216).[24]

Institutional Responses to Climate Change

Institutional responses to climate change are often slower to pick up on the multiple voices and views of the Pacific countries and their populations, preferring to approach the issues as a single global or regional challenge. Fragmentation and lack of cohesion in the region can also be a challenge (Goulding 2016: 191) especially as countries focus on their own needs (often very diverse within their own countries). There is some resistance to regional programmes with countries fearing the loss of a national voice, or less access to funding and training programmes, but there is also growing recognition that diverse impacts need to be dealt with through a diversity of responses, but more particularly, taking into account indigenous voices.

SECRETARIAT OF THE PACIFIC REGIONAL ENVIRONMENT PROGRAMME (SPREP)[25]

SPREP was created in 1982 at an intergovernmental conference on the Human Environment, held in Rarotonga, Cook Islands. The conference included all Pacific nations plus the colonial governments of various territories—New Zealand, Australia, United States, United Kingdom and France. Initially SPREP was modelled very closely on the UN Environment Programme's (UNEP) Regional Seas model.[26] UNEP was a strong supporter of SPREP, and in the early days channelled a great deal of funding through the programme. SPREP started as part of the South Pacific Commission and other agencies involved in its coordination included the South Pacific Forum (now the Pacific Island Forum [PIF], previously known as the Bureau for Economic Cooperation), the United Nations Economic and Social Commission for Asia and the Pacific (ESCAP) and UNEP. These organizations guided the operations of SPREP until 1988.

In the early days of SPREP governments and regional organizations tended to divide 'the environment' up into segments such as forestry, water and mangroves, and there was very little integration or attention given to population, urbanization, economic change and development. The meetings of SPREP were attended by high-level government officials and SPREP staff and scientists could attend only as observers. It took at least a decade before NGO delegates were tolerated, or permitted to address the meetings. Several universities in the region established a body aimed at giving technical advice on SPREP and other environmental projects,[27] but from early on in the Regional Environment Programme there appeared to be little recognition of the broad and integrated nature of environment and human development.[28]

Recent global changes and attitudes have of course had their impact on SPREP. Well before the United Nations Conference on Environment and Development (UNCED) meetings in Rio in 1992, the World Commission on Environment and Development had served notice to the world that economic policies had to take environment issues into account if there was to be a sustainable future. Pacific nations responded well to this and their reports to UNCED; *The Pacific Way*, and *Environment and Development*[29] placed a great deal of emphasis on human and economic resources as they relate to environmental issues, for example, poverty, uneven distribution of resources, value of forests logged and population growth. SPREP member countries also developed the NEMS where countries not only identified major environmental issues but also reviewed legislation, implemented environmental training and strengthened capabilities needed to reach environment goals and strengthen community awareness. The NEMS were to be practical, owned and developed by the countries themselves working in partnership with governments, communities and other relevant bodies. Unfortunately, the NEMS also locked countries into funding from SPREP based on the priorities identified in the reports, leaving little room for flexibility should new environmental issues emerge.

THE ALLIANCE OF SMALL ISLANDS STATES

An important and enduring outcome of the Rio meeting has been the development of the Alliance of Small Island States (AOSIS), a grouping of small island countries especially vulnerable to environmental disasters and of course climate change and sea-level rise. The grouping included Pacific nations, and those from the Caribbean, Maldives, Mauritius and Malta. They were very active at UNCED in Rio and called for a conference especially for small island nations. In 1994, the first of these SIDS meetings, the Global Conference on the Sustainable Development of Small Island Developing States, was held in Barbados, West Indies.

Although the aims of Barbados seemed initially to be similar to those of Rio (socio-economic development, vulnerabilities, the preparation of actions and policies relating to environment and development planning, and improving the capacities of small islands populations), the focus was on the very small size and vulnerability of such nations to any major environmental change, in particular that brought about by climate. A programme of action for the Sustainable Development of SIDS was developed and the results expressed in the Barbados declaration now underpin SIDS meetings. AOSIS countries were very active at world climate change conferences and the creation of the Pacific Small Islands Developing States (PSIDS) network has enabled a stronger Pacific voice especially in relation to the development of the United Nations Framework Convention on Climate Change (UNFCCC)[30] but also to regional and country involvement in the G77.[31]

DONORS AND THE PACIFIC NATIONS

Pacific countries are the recipients of a great deal of donor attention, as is the case for most other developing countries. Under a range of global environmental conventions, donors give support in areas such as the Convention of Biological Diversity, the Framework Convention on Climate Change, the Convention on the Law of the Sea, Convention on International Trade in Endangered Species of Wild Fauna and Flora (CITES) (trade in endangered species), the Basel Convention on the Control of Trans-boundary Movements of Hazardous Wastes and Their Disposal, the London Dumping Convention and many more.[32] Some of this support is with individual countries such as Fiji, the Marshall Islands, Kiribati and the Federated States of Micronesia, in the development of Biodiversity Strategy Action Plans, and some is at the community level through Small Grants from the Global Environment Facility. Commitments to global conventions can become a burden for small countries with limited capacity, but most take their commitments very seriously. There have been cases of countries ratifying conventions in order to access donor funding, but now the significance of such commitments is obvious. For example, in two key areas, almost 20 years apart, the Republic of Fiji has put global concerns for the environment into sensible, community-level practice. Firstly, in 2000, responding to the Convention on Biological Diversity in devising plans for protection and capacity-building, the country's Biodiversity Strategy Action Plan had some very positive outcomes. A biodiversity warden scheme was established in rural villages as an initiative of those representing traditional systems and has continued to be successful with very strong community participation in many aspects of conserving biodiversity. Then, in 2014, in preparation for inundation of coastal villages due to sea-level rise, a 'Fiji Relocation Guideline' was developed in partnership with the government, affected communities, different line ministries, NGOs and scientists, in preparation for the probable movement of at least 30 villages. Such advance planning not only makes sure the relocation of people from their homes and land is well coordinated and planned (and thus more likely to be accepted), it also indicates that the threats of sea-level rise are being taken seriously, and are dealt with in partnerships and not simply ill-prepared reactions at the time of inundation.

SPIRITUALITY, THE CHURCH AND CLIMATE CHANGE

In the 1960s Clarence Glacken wrote in depth[33] on the relationships of peoples and their environments, and the influence of climate on temperament, culture and the moral and social nature of human beings. He challenged the views of various writers, explorers and philosophers and was especially scathing of Western notions of human domination of the environment, harmonious relations of peoples with their environments and the false separation between nature and art. In Glacken's opus, custom, tradition and religious beliefs were demonstrated to be far more complex than Western thought had ever considered. Most people of the Pacific understand this but it has not prevented the romanticizing of early islanders and their environments. Clarke has also reminded us of the vast amount of traditional knowledge that *'has been or is being recorded, but in a sporadic way, scattered over hundreds of years and thousands of publications based on observations by geographers, anthropologists, biologists, agriculturalists, administrators, and early European explorers, artists, and natural scientists'*[34] and of course by Pacific Islanders themselves.[35] It is more widely understood these days that traditional knowledge *'can strengthen and enrich modern science and resource management'*, through the understanding of the holism of landscape and even an understanding and respect for beauty.[36] Such understanding and appreciation must surely include religious and spiritual beliefs, for example, pre-Christian belief in gods and ancestral (and non-ancestral) spirits to whom offerings can be made in return for good fortune, for example, with harvests and climate.[37]

In the twenty-first century, although it is understood that not all Pacific communities continue to live in harmony with their environments[38] due to the many competing factors of overseas education, urbanization and sheer greed, there is nevertheless a very strong role for religion, spirituality and traditional belief systems. Walter and Hamilton,[39] working within communities in Solomon Islands, have 'developed a cultural landscape approach that involves the construction of a conceptual model of environment that reflects the indigenous perceptions of landscape'. They have ensured the model includes 'cultural, ideational, and spiritual values' as well as more organizationally traditional-conservation methods. The model developed out of reactions by Solomon Islanders in coastal areas of Isabel Province[40] where people were far more interested in 'conserving cultural heritage sites than any other ecosystem resources'. There was much more enthusiasm for the adoption of conservation programmes based on cultural landscape models that recognized indigenous values.[41] Essentially what Walter and Hamilton recognized through their many years of living with these communities, was that 'ideological and spiritual values are key drivers in structuring the relationship between a society and the environment in which its members live out their lives'. They recommended that, for example, the Ecosystem Services Framework adopted by the Millennium Ecosystem Assessment Board of the United Nations (considered a key influential paradigm within the international conservation movement) must stress the contribution of aesthetic and spiritual values of ecosystems that provide community comfort and security.[42] Walter and Hamilton and the people of Isabel are not alone. Formal Christian churches in the Pacific are having a great deal to say, and some are also working very closely with communities across the region to present the voices of the people, not only from a purely Christian viewpoint, but encompassing and recognizing traditional belief systems.[43]

FORMAL RELIGION

Although conventions, meetings and treaties are now careful to refer to 'civil society' and rarely make specific mention of formal religion, there are nevertheless many examples of the Christian churches speaking out on climate change. Caritas, the Catholic Bishops Conference Agency for Justice, Peace and Development is prominent in giving voice to local and global issues. Caritas Oceania (an organization that has long openly discussed environment and climate change as a challenge to the peoples of the Pacific), published in 2014 a collection of short reports and interviews aimed at giving people a public voice on environmental issues such as climate, nuclear testing and logging. The collection is heartfelt and challenging, and demonstrates respect for formal (recent) religious traditions as well as traditional wisdom and knowledge. Pope John Paul the Second spoke about the peoples of Oceania (Ecclesia in Oceania, 2001) recognizing the need for better health of the Pacific Ocean. A comment from David Rauna of Solomon Islands, on the occasion of the devastating 2014 floods, perhaps sums up the understanding, (and the need for improved comprehension) of the impacts of climate change: 'the effect of the waves going inland was clearly visible. ... People had their gardens eroded...', 'If only people could be convinced on the effect of this climate change ... I think we will move to higher ground'.[44]

The Anglican Church, although small in Oceania (but most important in Solomons), is globally active through its Eco-Bishops Conference, a network which places great faith in learning from indigenous peoples. Recently, the Eco-Bishops have been discussing climate change as a way of thinking more deeply about what modifications people can make in their lifestyles to achieve climate justice.

One of the larger Christian denominations in the Pacific, the Methodist Church, tends to talk of spirituality as part of human responsibility to live in harmony with the environment[45] without specifically addressing issues of climate change but, like other churches, has been active in Agenda 21 in the 1990s, and more recently in the global climate change meetings. This approach may reflect a recent release from the Pacific Media Centre that claimed: 'Pacific culture, spirituality limit climate change news because ... people are linked to their land in a cultural and spiritual way which makes messages about their changing environment hard to receive.'[46] How these aspects achieve such an outcome was not explored and may reflect an unwillingness by journalists to probe more deeply, but the statement appears to be in contradiction to earlier comments by the formal churches, which, in more recent times at least, seem to be more open to building upon people's traditional knowledge and beliefs.

DEALING WITH 'THE ENVIRONMENT' IN THE CLIMATE CHANGE DECADES

Now, in the years of the UNFCCC Conference of the Parties and various 'Climate Deals' meetings, where it is sometimes unclear what exactly has been agreed and, most importantly, what will be the outcomes for the small island states of the Pacific, it is useful to look at how Pacific countries and communities themselves respond to global action.[47] Climate change negotiators across the Pacific region, including both government officials and people working in NGOs and community projects, contribute regularly to an on-line forum (Climate Change and Development Community) aimed at sharing information and ensuring those involved do the best with the resources available. Discussions revolve around how to better and more efficiently deal with the multiple meetings, tasks and reporting of global climate negotiations, as well as how best to implement effective projects on the ground. The forum has some useful queries and comments to make, for example, on the value of The University of the South Pacific's

(USP) Master of Science Degree on Climate Change, and suggests that a broader programme covering International Cooperation by the regional universities such as USP, University of Papua New Guinea, National University of Samoa, and Solomon Islands University would be very useful. People also actively engage with international NGOs such as 350.org, and participate in global on-line training and seminars.[48] For example, in 2017 more than 900 students enrolled in the on-line UNESCO supported/USP facilitated course entitled 'Climate Change and Pacific Islands' (www.mooc.usp.ac.fj).

Internal discussions of climate change actors aside, the politics of the Pacific region and various institutional arrangements seriously impact country responses at many levels,[49] but there is also much going on at the national and local community level. Most significant is the more outspoken view of some nations that they are no longer content to be told by New Zealand and Australia how to manage their affairs. Such views manifest themselves in reaction to the PIF being led at various times by an Australian or New Zealand diplomat,[50] and increasingly as Pacific nations form new links with major powers (such as China)[51]; the feeling is that New Zealand and Australia should take more of a 'back seat' in the region. For example, for many years there has been concern over where the Pacific fits into Australia's priorities. Recent budget cuts and the failure to send an Australian leader to any meeting of the forum for some years have been noticed.[52,53,54,55] The 2015 appointed Secretary General of the PIF, Meg Taylor of Papua New Guinea, has wide and long experience both in the region and as a senior World Bank official. She was therefore expected to be a strong and effective leader. Recently the PIF has been speaking more openly about the impacts of climate change and the annual meeting in Palau in 2014 ended with a call to action on the issue of global warming, with the 15-nation regional grouping saying there was 'no excuse not to act to curb climate change'.[56] Despite this, PIF remains aloof, with stronger Pacific voices in the regional groupings.

New regional arrangements, including institutional reform, as well as a push by Fiji to enhance the Pacific Islands Development Forum (PIDF) (largely funded by China), may mean a shift in the balance of power, at least among donor nations. If Fiji's leader, Bainimarama, gets his way, it is possible that Australia and NZ may no longer continue as full members of the PIF and could simply be development partners like Japan, China, Korea and the USA. Certainly, this will mean a greater role for China and India in the regional organizations, and probably more push for 'collective co-ordination and rationalization of aid'.[57] What this will mean for activities surrounding climate change is anyone's guess, particularly with the arrival of new players following recent Pacific disasters.[58]

The fact that PIDF excludes Australia and New Zealand is a lesson in listening, and the Suva Declaration on Climate Change demonstrated this very strongly, particularly in the emphasis on seeking local solutions. As Greg Fry, an Australian National University scholar, said: 'A Pacific Islands Forum with Australia and New Zealand as members is hampering the ability of the Pacific island states to defend their interests, and in the case of climate change policy, their very survival.'[59]

THE SIDS CONFERENCE OUTCOMES ON CLIMATE CHANGE

The text produced at the conclusion of the Samoa meeting in 2014 contained much of the rhetoric of earlier conferences, but what was particularly important was the major emphasis on climate change. Building on the foundation of sustainable development that had emerged from the Rio Declaration on Environment and Development (1992), 'the SIDS Accelerated Modalities of Action [S.A.M.O.A.] Pathway document ... recognizing that sustainable development necessarily includes attention to poverty eradication, the vulnerabilities of certain states, opportunities, economic development, the reduction of inequality and so on, firmly noted that sea-level rise and ... climate change pose a significant risk to SIDS and ... efforts to achieve sustainable development'.[60] Importance was placed on the serious issue of climate finance and the need for more emphasis on SIDS with limited capacity. Apart from noting the general failure in mitigation efforts across the globe, the document stressed the need to support the efforts of the SIDS, not only in increasing resilience and raising awareness through better communication, but also through more practical efforts of energy efficiency, disaster risk reduction and better management of the world's oceans, forests, biodiversity, water and sanitation, transport, health, education, gender, culture, social issues and food security. Significantly the inclusion of climate change and oceans as 'stand-alone sustainable development goals' is, according to Manoa, 'the PSIDS's most important success in advocacy to date'.[61] Such an inclusion and the all-encompassing list of themes has been seen in earlier documents, but what was notable this time was the acceptance of 'partnerships' with developing countries, in particular through climate change finance (para. 106), capacity-building and technology.[62]

It is to be hoped that the long list of comments on these issues of sustainable development do in fact translate into action. What was lacking in the document (it seems, although it most certainly came up in debate) was any obvious recognition of the knowledge of the peoples of the SIDS countries themselves and of what is being done at the local level.

COMMUNITY APPROACHES AND ADAPTATION

When particular actions and progress around the Pacific region are examined, it is clear, as will be demonstrated in later chapters, that Pacific Islanders have long managed (and damaged) their own environments, and local and community projects designed both to mitigate against and adapt to the ravages of climate change have had mixed results. What is also obvious is that there is a great deal of knowledge and activity held locally, and collectively these can lead to positive outcomes and lessons, not only for other Small Island States, both within the Pacific and beyond, but which can also be shared by the other Pacific cousins, particularly those of Aotearoa New Zealand.

NOTES

1. Many have challenged the environmental deterministic and other economic rationalist approaches to the island Pacific. Most famously the late 'Epeli Hau'ofa critiqued such views in 'Our Sea of Islands' where he provided an alternative view or vision of the Pacific which involved seeing the ocean as an integrated part of Pacific lives.
2. Carew-Reid (1989).
3. Now known as the Secretariat of the Pacific Community, the SPC, based in Noumea, New Caledonia, was created in 1947 by six major powers who (at that time) had political interests in the Pacific region—Australia, New Zealand, France, the United Kingdom, United States and Netherlands. The rationale for the establishment of the SPC was largely to secure Western political and military interests in the Pacific, but there was very little open debate, especially of nuclear testing. It could be argued that the constraints of debate in the SPC ultimately led to the establishment of the South Pacific Forum (PIF). Today the SPC is largely concerned with issues and themes such as economic development, fisheries and aquaculture, geoscience, health, social development, gender issues and statistics for development. It also has climate change projects.
4. Carew-Reid (1989).
5. Bryant-Tokalau (1994).

6. Supported by the Secretariat of the Pacific Regional Environment Programme (SPREP) and funded by Australian Aid.

7. Capacity 21 refers to some of the principles of Agenda 21 which were agreed to at the United Nations Conference on Environment and Development (UNCED) held in 1992. These principles included: environmental protection is an integral part of the development process; the need for all citizens to have access to information and other opportunities necessary for them to participate in decisions about environmental management; effective environmental legislation and standards; environmental costs to be internalized and economic instruments developed to facilitate that; the engagement of women in all efforts to achieve sustainable development; and special recognition given to indigenous people, their knowledge and traditional practices, enabling them to participate in the process towards sustainable development.

8. And, it should be said, the large numbers of consultants and international agencies which have worked with and for those countries. In recent years however, more local consultants have been involved in translating concerns into reports for bi- and multilateral donors and the Development Banks.

9. The two-week Earth Summit adopted *Agenda 21*, a blueprint for action to achieve sustainable development worldwide. Agenda 21, although weakened by compromise and negotiation, is considered to be one of the most comprehensive and potentially effective programmes of action ever sanctioned by the international community. Indeed, the Earth Summit influenced all subsequent United Nations conferences by placing emphasis on the integral relationship between environment and development, in essence support for socio-economic development while protecting the environment from deterioration and a partnership between the developing and more industrialized countries in order to ensure the planet's healthy future.

10. The Republic of Kiribati for example, with three island groups, spread across an ocean area of 35 million square kilometres does not have the resources to control poaching fishing fleets.

11. The Barbados meeting resulted in the Barbados Plan of Action, and Mauritius in the Mauritius Strategy, both of which had as a broad objective, the sustainable development of SIDS.

12. http://news.yahoo.com/climate-focus-un-small-islands-summit-samoa-020329795.html?soc_src=mediacontentstory.

13. http://news.yahoo.com/climate-focus-un-small-islands-summit-samoa-020329795.html?soc_src=mediacontentstory.

14. Goulding (2016) discusses the complexity of negotiating Pacific climate diplomacy and looks at the role, and success, or otherwise of regional institutions in trying to bring about a coordinated Pacific response in climate negotiations.

15. The Quaternary period is generally taken to mean the most recent 2.6 million years which includes the present day. The period was faced with dramatic climate changes, much species extinction and the rise of humans.
16. Nunn (1991), p. 5.
17. Nunn, p. 50.
18. ASPEI was the Association of South Pacific Environmental Institutions which was established out of UNEP (the UN Environment Programme) in the 1980s and tasked with looking at potential impacts of Climate Change.
19. ASPEI produced the 1988 report 'Potential Impacts of Greenhouse Gas generated climate change and projected sea-level rise on Pacific Island states of the SPREP region.
20. ASPEI produced the 1988 report 'Potential Impacts of Greenhouse Gas Generated Climatic Change and Projected Sea Level Rise on Pacific Islands States of the SPREP Region'.
21. Barnett and Waters (2016) do provide a more nuanced representation of vulnerability to climate change.
22. Bell, Taylor, Amos, and Andrew (2016).
23. George Carter (2015) Establishing a Pacific voice in the climate change negotiations.
24. Indigenous voices on Pacific climate change have become much more apparent in the past five years. Writers such as George Carter (2015) and Tammy Tabe (2016) writing both on regional politics and specific examples—such as i-Kiribati migration, are adding more layers to climate change responses.
25. Now called the Secretariat of the Pacific Regional Environment Programme, the South Pacific Regional Environment Programme was established in 1982, initially under the then South Pacific Commission in New Caledonia until it became an independent intergovernmental organization, based in Samoa in 1992. It achieved full autonomy in 1994. Initially the SPREP work programme and budget were submitted to and approved by the (then) South Pacific Commission, but since autonomy, and under the SPREP Agreement the organization is independent, guided by its member states. The work programme is focused around the activity areas of climate change and sea-level rise, protected areas and species, natural resources management, education, renewable energy, pollution control, biodiversity, coastal and watershed management and pesticides. South and Veitayaki (1999). Accessible at http://ncdsnet.anu.edu.au.
26. UNEP had regional seas programmes in the Mediterranean, Caribbean, Africa, Asia. The model was developed to find a way of multiple countries bordering the same ocean to develop a manageable plan for sustainably managing their ocean resources.

27. ASPEI, the Association of South Pacific Environmental Institutions, was chaired at various times by staff of the University of Papua New Guinea, and The University of the South Pacific.
28. Bryant (1994).
29. SPREP (1992).
30. Barnett and Campbell (2010), p. 101.
31. Goulding (2016: 196) also comments on the important roles played by PICs in major global alliances
32. Bryant-Tokalau (2008).
33. Glacken (1967).
34. Clarke (1990).
35. Gegeo (1998), Carter (2015), Tabe (2016).
36. Clarke (1990), p. 248.
37. Rudiak-Gould (2013).
38. Clarke (1990).
39. Walter and Hamilton (2014).
40. Isabel province in Solomon Islands is also known as Ysabel.
41. Walter and Hamilton (2014).
42. Walter and Hamilton (2014).
43. It is noted that at the time of writing this book, Patrick Nunn, author of 'The Hand of God' has more recently written on why secular climate change projects fail, and on the role of spirituality in climate change adaptation.
44. Caritas (2014).
45. Bhagwan (2014).
46. www.pmc.aut.ac, 12 March 2015.
47. People do comment in on-line forums, such as the Climate Change and Development Community (www.ccd-pc@solutionexchange-un.net) that despite the importance of climate negotiations, keeping PIC delegates engaged in the international negotiations is a challenge. Discussion currently surrounds ways of spreading the load, so each of the 14 countries can take the lead to coordinate various issues of interest and importance.
48. ccd-pc@solutionexchange-un.net, 6 November 2014.
49. Micronesian subregional politics have not been discussed in depth here, but Gallen (2016, pp. 176–188) does point out that there are different allegiances within the area loosely named Micronesia. The same could be said of other parts of the region, belying the views that all are in agreement over key issues.
50. Sometimes in the past this has occurred after a period of concern over certain forum leaders whereby a more 'neutral' leader might be a safer pair of hands.

51. In November 2014, the leaders of China and India met in a summit with Pacific countries. China discussed, for example, the feasibility and implementation of the US $2 billion concessionary loan for infrastructure development in the eight PICs. Climate change support was high on the agenda, although details of any specific climate change projects and how the largely infrastructural projects will be funded and loans repaid, for example through the Exim Bank's concessional loans is yet to be outlined. 'China keen to assist infrastructure development and climate change in the Pacific' By Online Editor 2:59 pm GMT+12, 22/11/2014, Fiji. http://www.pina.com.fj/index.php?p=pacnews&m.
52. Iati (2010).
53. Powles (2010).
54. MacLellan (2015a).
55. It should be recognized however that Australian Aid funds the Pacific Leaders Programme and as part of that gave support (2012/2013) to 'Future Climate Leaders' (FCLP2) working with the University of the South Pacific's Centre for Environment and Sustainable Development. The aim of the project was to enhance 'skills and knowledge of climate change staff and students as well as Pacific Islanders at the community level' (Australian Aid/USP 2013). With cuts in Australian Aid in 2014, some projects in the overall leaders programme (such as awards) have currently been delayed but the PLP continues in 2015.
56. http://www.biznews.com/green/2014/07/31/climate-change-pacific-islands-likely-disappear.
57. https://narseyonfiji.wordpress.com/2014/11/13/bainimarama-and-the-forum-a-storm-in-a-calm-ocean-devpolicy-blog-11-nov-2014/.
58. Fry and Tarte (2016) came to a similar conclusion with the main concern being what the pressure on and shifting of allegiances will mean for Pacific peoples.
59. Fry (2015), p. 1.
60. para. 11, www.sids2014.org/samoapathway.
61. Manoa (2016: 96).
62. para. 105, www.sids2014.org/samoapathway.

CHAPTER 3

Adaptation to Climate Change in the Pacific Islands: Theory, Dreams, Practice and Reality

Abstract In this chapter cases are presented on how people respond to climate change. The modern artificial island response of donors is presented, but with the reminder that Pacific Islanders have themselves long known about reclaiming and building up land in order to mitigate against climate change. The ways that people respond to extreme weather such as in cyclones are also illustrated with examples of traditional methods of preparedness such as storage and planting and recognizing signs of nature. Relocation, long practised by Pacific peoples, is also discussed, along with some of the pitfalls of moving into other countries and communities.

Keywords Adaptation • Artificial islands • Relocation • Land purchases

Women's groups are saying to me and to those that are listening, very strongly, 'we aren't just vulnerable' or 'we aren't vulnerable, or we don't want you to portray us as a vulnerable group, there are vulnerabilities, [and] vulnerabilities can be exacerbated but women and other groups are strong, they're resilient'. A. Miller (2016). http://www.radionz.co.nz/international/pacific-news/311609/pacific-women%27s-voices-crucial-for-disaster-management

© The Author(s) 2018
J. Bryant-Tokalau, *Indigenous Pacific Approaches to Climate Change*, Palgrave Studies in Disaster Anthropology,
https://doi.org/10.1007/978-3-319-78399-4_3

INTRODUCTION

Adaptation to climate change is about adjustments that people need to make to limit the impacts of climate change, but it can also include looking for opportunities there might be in a changing climate. Adaptation is urgent—it is not only a future necessity but also a current reality. To adapt though, resources must be available. These are financial and technological, but they can also be knowledge based, including traditional knowledge.[1] In this chapter, it is argued that traditional forms of adaptation to climate change are often overlooked and that adaptive responses may have the best likelihood of success if both traditional and modern approaches are combined. The range of 'hard' options, both traditional and high technology, is outlined as contrasting responses to similar issues, but with very different outcomes. Further adaptations examined include relocation and wise environmental management, drawing on traditional ecological knowledge. Each form has difficulties and each targets different populations. What becomes clear from the case studies is that only where Pacific Islanders themselves have agency over responses to climate change, will there be successful outcomes.

Science has warned for much of the past two decades of the urgency for nations and their communities to adapt to protect people from the ravages of inundation, more intense storms and the loss of drinking water and gardening land. How adaptation is implemented is complex, however, with people not always accepting the science, or being unable to respond effectively due to lack of funds or adequate support. Realistically, adaptation has to occur across all sectors of society but is most likely to be successful where local communities are onboard.[2] It is also crucial to recognize that the 'adaptive capacity' of those impacted by climate change can be found in relationships, participation of traditional leaders and churches, and links throughout governments, communities and organizations. Adaptation is not solely about money or high technology solutions. Many successful adaptations come from using what already exists, and adapting traditional technologies. Societies know that they adapt or perish, but sometimes what is already known is forgotten. This is now the time to use such knowledge and to do something to avoid the inevitable 'slow disaster' of a 'disaster in the making'.[3]

There are many modern approaches to adaptation including coastal protection projects (sea walls, dredging), water supply provision (desalinization, better capture and storage), coastal planting, alternative energy and other 'hard' infrastructural projects. There has also been a flurry of environmental/climate change education, capacity-building and other 'soft' projects. These have been implemented by a range of donors and institutions from the Global Environment Facility (GEF) administered through the World Bank, UNDP, UNEP or the ADB, major global NGOs, local NGOs, educational institutions and bilateral donors. Not all have been successful with issues of overlap, poor management, funding crises and donor competition[4,5] reducing their impact. Few of these projects incorporate traditional ecological knowledge in their design.

Several approaches to adaptation will be examined in some detail. The 'hard' approach of constructing artificial islands is presented as a contemporary issue of donor and private sector assistance offering a form of adaptation, while neglecting local knowledge of adaptation by construction. For example, the construction of high seawalls to protect small islands can suffer from breaching, destruction and decay and may make some communities more vulnerable. Other forms of adaptation include relocation to other countries or areas. Such adaptations are outlined here, detailing both the pitfalls and the possibilities of these approaches. Traditional forms of adaptation and their continuation today should never be overlooked and the theme throughout this chapter is the urgency of working towards community-level adaptation if there is to be any chance of success in the short time frame now available.

'Hard' Options: Artificial Islands and Construction

In 2010, the world became aware of millions of kilos of waste afloat in the north Pacific gyre.[6] As noted elsewhere,[7] the recognition of the existence of such vast areas of plastics and other debris led to a range of excited responses from the international community. There were the 'plastic fantastic' plans for creating islands from plastics, essentially 'recycling' them for use and at the same time solving issues of overcrowding and threats from climate change. These approaches have become known as the 'Dubai strategy' of building artificial celebrity islands (with Dutch engineering), and also addressing more serious plans for small Pacific nations threatened by inundation from rising sea levels.[8] Some commentators even envisaged Pacific Islanders behaving like 'nomads' floating across the oceans.[9,10]

At the time, there was much debate about the sheer impracticality of artificial constructions, with the massive engineering feats of Denmark, Hong Kong, Azerbaijan and 'The World' in Mexico cited as examples.[11]

People have always found ways to create more living space when land is limited or waste disposal becomes a problem for settlement. In the 1980s there was a plan for California to ship domestic waste to Majuro, Republic of the Marshall Islands, in order (it was said) to extend and raise the level for settlement.[12] This did not happen, but the use of the Pacific gyre might be viewed in a similar light. Such externally funded projects often fail because they rarely take into account local existing knowledge and may be more to do with donors carving themselves a stake in the Pacific than genuine attempts to find successful outcomes. More successful Pacific examples of 'artificial' construction involving practical, local solutions include the extension of fishing communities such as in Koki and Hanuabada in Papua New Guinea, and where land is threatened by rising seas or enemy invaders such as in Bau in Fiji. Bau, and indeed the wider Rewa delta, also suffered from a series of apocalyptic events in 1840 with flooding following a major storm.[13] Although there is little comment on the link between having constructed additional land and the survival of the people of Bau, it is likely that it helped. Some of these artificially constructed islands have survived for centuries and others, such as 'The World' in Mexico, drifted away and were destroyed by a hurricane.[14]

In more recent times, coastal tourist resorts throughout the Pacific are most popular (despite, apparently the risks) and, if not built within a few metres of the ocean, are commonly built either out over the lagoon (such as Momi Bay in Fiji), or on reclaimed land, most often filled-in coastal mangrove areas such as Vulani and Denarau (also in Fiji).[15] Essentially these are 'artificial' islands, although not built with protection from rising sea levels in mind.[16]

Pacific tourist resorts are not the only examples of artificially created oceanic living space. In 2009 when Pacific leaders proposed that such islands could be a way of combating climate change and rising sea levels, it is possible that some may have been drawing on widely held local knowledge and precedents. In response to the leaders' call, Japan provided well-engineered plans for circular islands with enough space for the populations with farmland and for businesses. These would be located near the equator and would retain access to 'their own fishing grounds … and nation'.[17] Japan proposed that these islands could be ready for occupancy by 2025. Although such ideas appear quite fanciful and have been followed by even more elaborate plans, some Pacific leaders have shown interest, viewing such constructions as 'the best hope' for countries threatened with the loss of their homelands, as well as a better alternative to mass migration to a new, colder and possibly more hostile country.[18] There is limited academic analysis of the long-term outcomes of such constructions except for concerns over legal issues relating to ownership and the law of the sea,[19] but it is important that the nations possibly affected remain alert to potential outcomes.

In 2013, for example, Japanese company Shimizu Corp., announced its plans for a city that floats on vast 'lily-pads' on the surface of the Pacific.[20] The futurist design involved 'island(s) nearly two miles across, with a central tower rising half a mile to form a "city in the sky". The tower will house residential units for 30,000 people and space for offices, services and shops.' The company claims the city will be self-sufficient in food with the central shaft of the island being used to grow vegetables and fruit. The flat base of the island, which will be tethered to the ocean floor far below, will have a residential zone for another 10,000 inhabitants, along with forests, beaches and arable land, as well as port facilities.[21]

Further, Shimizu claims that individual units can be connected to form floating 'cities' of up to 100,000 people. 'The idea behind the Green Float project was first as a solution to the problem of a rapidly growing human population or as a city that would be immune to earthquakes and tsunami,' said Masayuki Takeuchi, the head of the scheme. 'But we quickly realized that it could save islands from rising sea levels. We are still at the planning stage, of course, but we believe this is a feasible project'.[22]

The then President of Kiribati, Anote Tong, quickly held discussions with Shimizu to consider buying these floating islands in order to relocate people. As can be seen by other efforts of Kiribati, and a global understanding that the people of Kiribati are among the first seriously endangered by sea-level rise, it is not surprising that such a radical (and perhaps fanciful) solution was considered. The UNFCCC and scientists generally recognize that global commitments to emission reductions as one step in reducing the impact of climate change will not help Kiribati (or indeed any other small island state that produces few greenhouse gas (GHG) emissions) and it is understood by Kiribati that: 'The momentum of what['s] already in the atmosphere will ensure that sea levels rise above our islands. We will be totally devastated ... [these are] concerns we jointly share about the increasing severity of the challenges facing our people today from climate change and the slow pace of global action to address them.'[23]

Shimizu's proposed 'installations'[24] are not really islands in that they will not be lodged on the seabed. They will more likely be anchored through moorings,[25] with heaped material on the sea floor. Not only do these plans have many possibilities of physical and technological failure (especially in the face of intense weather patterns such as increasingly powerful cyclones), and massive expense, they are also most probably being contrived for political purposes. Aid, relationships and political dominance of the regional seas by aid donors can be uppermost in the offers of 'assistance'.

Although, as can be seen in the agreement signed in 2017 between the Seasteading Institute in California and the government of French Polynesia, the Pacific's first seastead community may be about to be built near Tahiti. This community will be built entirely within French Polynesian territorial waters (unlike the usual international seas approach of Seastead). Once arrangements such as local community government approval (and that of France), benefits to the local economy and environmental protection are sorted then it is anticipated that the project will go ahead. It is not at all clear what physical form the project will take but aims include no fossil fuels and no destructive use of the seas[26] but with more than one thousand donors having contributed over $2.5 billion already, the plans to provide floating islands to 'people threatened by rising sea levels'[27] appear to be well underway. How far this island will benefit the local, less well-off Tahitians is unclear however and what genuine community consultation has been carried out is unknown. It may be that, as was the case with plans for floating islands in Kiribati, there has been no consultation with the people themselves about what they want to do. In Kiribati though, there was little recognition of the fact that most atoll and coastal Pacific people are sea faring rather than gardeners. In French Polynesia, the floating platforms that will house people do seem to be giving more recognition to the practicalities and the urgency of preparing for sea-level rise. At first sight, highly technical solutions to climate change through the expensive, privately funded construction of artificial or floating islands may appear to be fanciful and of no relevance to a discussion of Pacific Island climate change adaptation. But examined alongside what Pacific Islanders have themselves carried out over many centuries they may well demonstrate what is possible, if those impacted by the changes, can be part of the process. Traditional Ecological Knowledge, understood and accepted as valid, may offer some workable solutions.

Pacific Island Responses

Pacific Island populations are well aware of problems associated with climate change and are looking for alternatives, as well as building upon and utilizing long-held traditions and local knowledge. What kind of alternatives may depend on who is offering, what the costs are and what will be received in return. In these circumstances, few corporations, donors or the governments themselves take time to involve communities in the decision-making process. In the case of Kiribati there is little agency by the people who will supposedly be settled on the proposed artificial islands. There also seems to be a lacuna in knowledge of past efforts and people's capabilities. In Vanuatu on the other hand, as can be seen in the wake of the devastating impact of Cyclone Pam in March 2015, traditional approaches to preparedness, although small when faced with such powerful winds, may well prove to be the saviour of many on the small outer islands where food security and water supply might have been destroyed. Peoples of the Pacific have always had to adapt to environmental change and the threats of climate change are an extension of what has always faced atoll and coastal dwellers—overcrowding, brackish water, waste disposal, health issues and the problem of finance.

Pacific Island Responses: History of Artificial Islands

Long before Japanese and Californian offers of artificial islands, people have been constructing islands, building up land and generally adapting to their environment.[28] Of critical importance was that the people of 'need[ed] to combine scientific knowledge of climate change with traditional responses to historical change, including the stonework tradition and the cultural determination to resist undesired coastal change.[29]

In order to be sustainable, constructions should mimic the local surroundings. There are a number of cases in the Pacific, for example, the vulnerable atoll of Kapingamarangi (Pohnpei, Federated States of Micronesia) where people have manipulated naturally accumulating sandscapes to make them liveable, in the form of Touhou islet.[30] But not all artificial constructions are successful. Poorly designed modern sea walls and increased coastal erosion are well known around the Pacific[31] and the mining of coral on Tarawa by unemployed locals selling to their neighbours for cement making, is undermining efforts to keep the coast-line secure.[32] But successful cases of artificial construction are worth noting in the context of modern artificial islands in the Pacific. Some, such as Nan Madol, the ancient centre (AD 800 to 1500) of the Sau Deleur Dynasty of around 25,000 people in Madolenihmw in Pohnpei, Federated States of Micronesia, must have required enormous community cooperation, possibly even the use of slaves, or very powerful leaders with numerous willing commoners to carry out the work required to build huge residential basalt temples and canals with some blocks weighing up to half a ton.[33,34,35] Far-reaching connections throughout the Pacific would also have been necessary, both to access the stones required and to maintain the power of the dynasty.

Such societal organization would also have played a major part in maintaining structure and order prior to and following major disasters. D'Arcy discusses responses to devastating typhoons in Micronesia, for example. Collective responses to food shortages such as equal sharing of remaining food, alternative approaches to fisheries, where canoes were destroyed, were made possible under the direction of the chiefs, and new political links forged throughout the islands[36] but of course long-term survival depended on the extent of the damage. In the eighteenth century a typhoon resulted in the loss of a significant proportion of Kosrae's crops and population and according to D'Arcy many customs and traditions were also lost meaning societies had to be rebuilt, often through migration from neighbouring atolls[37] (Fig. 3.1).

LANGALANGA LAGOON, MALAITA, SOLOMON ISLANDS

On the western side of Malaita, where there is urgent need for additional land as large steep land masses often drop straight to the sea, the centuries-old (possibly as old as 1000 years)[38] small artificial islands (or islets) of the Langalanga and Tai lagoons are located. The origins of the islets have been widely debated by scholars over the past century.[39,40,41,42] Parsonson attributed the origins to the ravages of malaria, which may have necessitated migration by people from the bush of the main island to the clearer air of the lagoons. He also noted the artificial islets were most often constructed in shallow waters, with deeper water at one end enabling easy fishing from the houses.[43] The origins and continuing existence of these islands are likely to be more complex with a range of explanations such as the need for protection from enemies, the demand for an expansion of living space

Fig. 3.1   Nan Madol, Pohnpei. Source: Bryant-Tokalau photograph (2000)

for coastal people, the bartering of fish for root crops between the 'sol wara and bus people'[44] as well as protection from mosquito borne diseases. The people of Langalanga continue to be well known for their skills as boat builders, and as creators of traditional shell money. The houses built on the small artificial islets of Langalanga (or Akwalaafu) are, along with the beauty of the large natural lagoon today, viewed as a highlight of western Malaita for the growing tourist trade. More importantly however is the history of the many artificial islands.

The importance of Langalanga and other artificial islands is bound up with the way people conceptualize and appropriate their landscape to demonstrate the methods they use to remember and continue to use their history. Although Langalanga is now firmly in the 'tourist gaze', this is not the only reason for their continuing existence, despite Guo's claims that 'the colonial and post-colonial (tourist) gaze have highlighted artificial islands as a locus of interest'.[45] Other research, such as that by Hviding, demonstrates more complex reasons for the continuing existence of these islets, particularly institutionalized exchange between the lagoon and bush people and trading fish for root crops.[46] Such transactions have not diminished over time and indeed the communities are thriving and continuing as before[47] (Figs. 3.2 and 3.3).

The artificial islets of Langalanga have much wider significance than for Malaita alone. Of particular importance is the relationship between peoples of the very vast Pacific. The origins of most people living on Melanesia's artificial islands, point to their relationship with Polynesian and Micronesian peoples who frequently occupy the 'outliers' further afield in Melanesia such as Ontong Java, Trobriand Islands, Tikopia and Futuna.[48] This is evidenced by a 'strong Micro-Polynesian strain', and cultural affinities such as fishing methods (kite fishing, for example, as well as outrigger canoes), shell money and the building of houses on piles—but more particularly the design and decoration of the houses. Canoe burial, similar statuary and names originating from Tonga (Lau, Langalanga) as well as the ancient arts of stone building can also be found throughout Polynesia and Micronesia (Pohnpei, Tonga, Marquesas and Hawai'i).[49]

Fig. 3.2   Kokoifou artificial island, Lau Lagoon. Source: Sally Hundleby

Fig. 3.3   Lilisiana village, Langalanga, Malaita. Source: Sally Hundleby

The links between the artificial islands of Lau and Langalanga and climate change adaptation are unclear, and as described above, may well be a useful response to inundation despite having been built for other reasons. Of significance are the relationships that people forge and the knowledge shared where societies maintain strong links. Malaita has faced a number of significant traumas in the recent past both social and environmental, but according to Walter and Hamilton[50] 'indigenous communities tend to embrace a more complex and diverse set of environmental values than is often appreciated'.[51] Supporting Guo's view of landscape as memory, they discuss the encoding of cultural memories and history that is then used in socialization and social reproduction. They also point out that in Solomon Islands (and other coastal communities) shrines and other sacred sites are not simply viewed today as static reminders of the past, but are used and remembered in very modern transactions such as in logging or land deals. Indigenous writers of the Pacific constantly refer to the fact that cultural and physical notions of landscape are thus intertwined.[52] Such an understanding is very significant when trying to analyse community adaptations to climate change and provides a deeper understanding of why and how people move and respond to their landscapes. Although Langalanga did not develop as a response to climate change, and there were many reasons behind the development of the islets, the fact people have developed such islands and lived on them for many generations demonstrates the skills people have in creating new land, but also the continuing memories, which enable the past to continue in the current landscapes. Even if people move and construct new homes, memories of ancestral links, ancient voyaging and conflict remain in both stories and actions.[53]

These memories and abilities are also significant when looking at other forms of adaptation such as relocation and migration in response to inundation and/or loss of territory through climate change. Following are presented examples of traditional practices of adaptation that can save threatened peoples, and also traditional melded with modern institutional forms of adaptation. I will then examine cases of relocation in the contemporary age, in particular one planned mass migration to another country. These examples of adaptation will be used to demonstrate how governments sometimes fail to involve their own communities in decision-making and under-estimate the significance of people's cultural memory and wishes. This is interesting because despite wide media and scientific coverage on the global stage, even those tasked with helping their populations adapt to massive future environmental change sometimes fail to recognize the knowledge and wishes of people who have travelled far and wide to settle on their current islands.

FURTHER TRADITIONAL ADAPTATIONS

Although there are many traditional forms of adaptation to climate change, some are overlooked, especially in the wake of modern disasters. Traditional and modern approaches are often combined, leading to new adaptive responses. In Chap. 4 where responses to floods are examined, I outline how people prepared for Cyclone Val in 1975 by using, for example, time-honoured methods of crop preservation. The 2015 Cyclone Pam that hit the Pacific region, a 'monster' category five storm, annihilated much in its path, not only in Vanuatu, but also in small outlying islands of Solomon Islands. There were also major storm surges that inundated islands in Kiribati and Tuvalu. Global assistance was massive, but issues of distribution and access to islands and communities with no communications, airstrips or the means of travel made the delivery of assistance extraordinarily difficult. Additionally, the Vanuatu government was determined that aid distribution be fairly done and without the chaos that multiple donors can sometimes bring. Although many viewed this as dangerous and unfair, especially where people had no water or shelter, other stories quickly emerged. In a missive from Tanna, Tessa Newton Cain reported that:

> *For now, people will eat what can be salvaged from the storm. Some will have livestock that they can kill. ... in Middle Bush, Tanna, where all are safe (largely thanks to having endured the cyclone in daylight hours), people are making use of solar powered driers to convert manioc into flour before it rots. That will be a much-valued resource in the weeks and months to come. ... we are recovering. This is a resilient country populated by resourceful people and we are working together to get things back on track. There are many challenges ahead: logistical, political, economic and social. There is need for much assistance and we know it is on its way. When it gets here, it will find us already working hard in our island home.*[54]

Government announcements reminded the world that people could eat yams for up to a week after the event, but such comments may not have had wide application as saltwater inundation as well as the sheer force of the wind and rain caused havoc, including to root crops. Stories of people drinking seawater may also have been exaggerated as fresh water springs are common on the volcanic islands. Despite extreme hardship and loss of life, by the time aid reached isolated communities, most people had found ways to survive without starving. Before cyclones hit, people in Vanuatu and elsewhere prepare food for use after the devastation. Cyclone foods include the harvesting of crops such as banana, which are wrapped in banana leaves and buried to eat later.[55] Burying food supplies is widespread, and perhaps under recognized by outside agencies in the wake of such major tragedy. Other factors though, such as volcanic eruptions, have caused contamination of some root crops on Tanna. Disease is a much bigger problem, and the wider impact of such a storm, made worse by higher sea temperatures than usual, will have a long-term effect on livelihoods, education and health as rampant illnesses have taken their toll on even the most resilient populations. While used to regular environmental upheavals, Pacific societies may be viewed as being unused to such all-consuming disasters, such 'unanticipated powers of nature', but so too do they find ways to explain such destruction. In many societies, whether strongly Christian or retaining powerful spiritual or supernatural beliefs, disasters may be explained as happening for a reason. Angering the gods, or doing something unacceptable, even several generations before, can lead to severe consequences, yet be accepted as inevitable.[56] How people will respond in future, and whether more powerful forms of adaptation, incorporating more indigenous knowledge, yet also sometimes accepting what 'will be, will be' remains to be seen, but since people need to survive, new gardening, mangrove planting and building practices may be created and the immediate response may not always be one of flight, at least not from the higher Melanesian islands.[57]

Modern Alternatives: Relocation

Peoples of the Pacific have migrated, been relocated and resettled for all of their history in the vast world of Oceania. Much of this has been in response to environmental changes and challenges, but it has also come about through competition for resources, political changes and over-population. All migration comes with challenges both personal and social, as well as political and physical. With contemporary adaptation to climate change where the urgency is becoming greater, it is wise to take heed of historical pitfalls and challenges, and especially lessons of history and beliefs that can often be overlooked.

The history of relocation will not be repeated here. Several writers have commented on resettlement in great detail[58,59,60,61,62,63] pointing to the difficult decisions that migrants must make, the adjustments, the return home when things do not work out and, of course, the extreme challenges of access to land, resources, and employment, maintenance of language and other cultural practices. Not all migration is negative, and many migrants relocate very successfully; after all everyone is a migrant since that is how people arrived in the Pacific initially. Among the plethora of papers about Pacific migration there is also a warning voice—how many people have actually migrated due to climate change? To date there is little evidence,[64] but whether such movement is yet a necessary adaptive response to climate change or not, countries are seriously pushing relocation as a major response. Viewing migration as the only possible option can remove ways of utilizing traditional knowledge. It can be a risk when there is strength in TEK and the ability of people to adapt.

One case study presented here is of the Republic of Kiribati, a nation that has become well known on the world stage as a small, low-lying country in danger of inundation through climate change. The case is interesting as overseas migration is being promoted when in fact there are several options available within the many islands of Kiribati. The movement of people away from flood-prone areas to higher ground has always been practised and could be viewed as a 'risk reduction' strategy in the face of disasters.[65] Each country and even island have to be regarded as different. It is not easy for people to simply be 'transplanted' as demonstrated by Ursula Rakova's writings and energy in finding homes for her own people of the Carteret Islanders in Papua New Guinea to Bougainville Island.[66] Not only is Bougainville very different to the low-lying Carteret Islands both physically and culturally, it has also been through a lengthy war. The challenges faced by the people demonstrate the extreme upheaval of moving, not very far, in the face of rising sea levels, to end up in an area with different languages, beliefs and ways of survival. And of course, there are also cases of peoples moving to even more different environments by travelling internationally, often to cities. Kiribati is looking at many options, and one of these is to relocate to a rural area of Fiji.

RELOCATION: KIRIBATI AND FIJI FUTURE IMPERFECT

The Republic of Kiribati straddles the equator, north of Fiji and Tuvalu (Fig. 1.1). There are 33 atolls and reef islands and one raised coral island: Banaba.[67] There is a total land area of 800 square kilometres (km²) spread over 3.5 million km² of the Pacific Ocean. The permanent population is around 110,800,[68] and in 2015 over half lived on the atoll of Tarawa that includes the capital, South Tarawa that at just 15.8 km² in size has a population density of over 3000 people per km². Kiribati has a humid, tropical climate with a temperature range of less than one-degree C but warming has been observed in the past few decades, especially sea surface temperatures.[69] Sea-level rise since 1993 is between 1–4 mm per year (cf. global average of 3.2+ or −0.4 mm per year[70]). Extreme sea-level events, such as king tides, always occur during El Niño conditions.

Over a 2000- to 3000-year period between 3000 BC and 1300 AD, Kiribati was settled largely from Micronesia. There was also much interaction over the centuries between Kiribati and countries to the south, especially Tonga, Samoa and Fiji, often in the form of invading parties.

Although Kiribati is regarded as 'Micronesian', links with other nations have meant there are strong ties with Polynesian and Melanesian cultures. More recently (from the late 1930s) Kiribati people have been relocated both within to the Phoenix group of islands,[71] and 'without', to Solomon Islands, largely as a response to environmental change.[72]

Kiribati, along with Tuvalu, (its former colonial partner in the British-governed Gilbert and Ellis islands), is often held up as one of the global poster children of climate change. As an atoll nation, Kiribati already has a very limited supply of fresh water, caught in tanks from rainfall or sitting under the atolls in fresh water lenses. Rising sea levels and increasing inundation means salt water infiltrates groundwater, and there is contamination from raw sewage and from gravesites. The disposal of rubbish is a constant dilemma in atoll countries. Apart from household rubbish, solid waste such as old cars, oil, batteries, plastics, hospital waste is a difficult and expensive issue given that the only disposal solution is to remove it altogether from Kiribati.[73]

Additionally, coastal erosion has led to the abandonment of some villages, and the atolls of Tebua and Abaneua disappeared in the late 1990s.[74] Coral reefs are being degraded by ocean acidification and over-fishing, meaning the peoples of Kiribati, as with atolls everywhere, are not only losing their livelihoods but are also being threatened by greater coastal erosion as reefs can no longer protect the islands to the extent they did before. On land, the few crops that grow, like breadfruit and coconut, are so badly affected by saltwater inundation they will die within three months after remaining in salt water for two to three days.[75]

Some scientists do not believe that the islands will be completely uninhabitable in a short climatic period, citing the natural building up of reefs by 10 to 15 mm a year (faster than sea-level rise).[76,77] There is also some debate over the use of images of large waves crashing into houses as some photographs used are of king tides, or are the result of badly constructed seawalls, and because new causeways between islets have altered the behaviour of waves.[78] International scientists are not the only ones to offer alternative scenarios to the images of Kiribati 'drowning'. Naomi Biribo, a Geography graduate from the University of the South Pacific and a senior scientist in Tarawa, authored a report with Woodroffe which essentially said the widespread erosion and flooding facing South Tarawa is largely due to local human activity, such as overcrowding, coral reef mining and the construction of sea walls and causeways.[79]

Whatever the nature of climate change effects in the face of potential devastation, Kiribati has been considering a number of options. These range from migration within Kiribati itself,[80] or to Australia, New Zealand or other Pacific nations; doing nothing, adapting and staying put, and very expensive engineering solutions such as the construction of suitable sea walls or building up the islands. More recently, as noted above, the government has been party to discussions with Japanese companies, which seek to build high-technology artificial islands. All are fully aware that the reduction of emissions by the people of Kiribati will provide no solace in the global mission to reduce climate change effects since island nations contribute so little, and that the die is cast for global climate change. Kiribati has always been vocal at international summits and many 'solutions' are offered to ease their plight, though some have caused dissent in the communities.

The former President of Kiribati, H.E. Anote Tong, was always an outspoken commentator on climate change and worked tirelessly to combat what he viewed as an issue that 'threaten[ed] the survival of a number of countries within our region and beyond'.[81] In 2012 however, an i-Kiribati journalist, Taberannang Korauaba, argued that (former) President Tong's growing focus on climate change was centred on his close relationship with the foreign news media and, as a result, has increasingly portrayed the people of Kiribati as victims. His argument was that the victim approach further marginalized people's ability to learn about climate change. But he also noted there is no connection between what Tong has declared overseas and his government's 2008–2011 Development Plan. Kiribati was not united on climate change: 'Traditional, cultural and religious beliefs about land, environment and sea, and division among educated elites and political parties are some of the key barriers to communicating and receiving climate change stories.'[82] In the Government's second report to the UNFCCC,[83] for example, there is no discussion of major external migration or the construction of artificial islands. The report outlines adaptation strategies ranging from secure water supply, to internal relocation, providing a detailed and well-balanced examination of Kiribati's current situation.[84]

One illustration of differing views over how Kiribati should adapt to climate change can be found in the debates over the purchase of land in Fiji.

THE FIJI LAND PURCHASE

The Republic of Kiribati has purchased 6000 acres (around 23 km² or 2331.3 ha) of land in Natoavatu estate near Savusavu on the island of Vanua Levu (Fiji) to ensure 'food security as its own arable land is swamped by the rising sea'. President Tong said he bought the land so his people will have some high ground to go to when a rising sea makes his nation of 33 low-lying coral atolls unliveable. 'We would hope not to put everyone on [this] one piece of land, but if it became absolutely necessary, yes, we could do it' he told the Associated Press.[85] For years, Tong claimed in climate change conferences and in interviews that sea-level rise was already taking a heavy toll on his people, eroding beaches, destroying buildings and crops, forcing the evacuation of a village, and wiping out an entire island. His views were echoed by the NGO, Conservation International, on whose board Tong sat. The residents of 'Kiribati, where the effects of rising sea levels already are being felt, [are] on the front lines of climate change,' says its website.[86]

There is a great deal of discussion about this purchase with i-Kiribati scientists, civil servants and atoll dwellers in debate over the newly acquired Fiji land. FJ $8.7 million has been paid, taken from Kiribati's $600 million sovereign wealth fund, whose interest goes into the budget.[87] According to former President Tito, Kiribati paid four times more per acre than other buyers in the last few years. He believes that the price paid had been done solely for publicity purposes to highlight Tong's far-sightedness and how seriously he takes climate change.

Hermann and Kempf[88] refer to the land purchase as 'politics of hope' by the governments of Fiji and Kiribati, allowing people to develop 'imaginings of migration ... hoping that, in the event of an existential threat, this new land will allow them to preserve culture, nation, and identity over the long term'. But there is a much darker story surrounding this purchase, one that shows that the option of relocation in the face of climate change, although it may sometimes be the only option, can lead to tension and possibly no hope for some.[89]

Although government officials in both countries celebrated the purchase, noting the land is capable of accommodating 60–70,000 people, there are many problems. In Fiji, John Teaiwa, the former Secretary of Environment who is a Rabi islander, says the land makes no sense as a food source. The land is in the interior of Vanua Levu, a high, mountainous island, far from the atoll environment of Kiribati people. It consists largely of an abandoned coconut plantation and dense forest on steep hills.

Apart from difficulties in agriculture, there is another, equally serious story associated with this land. Two hundred and seventy Solomon Island descendants of 'black birded labourers' live on the former estate, invited there in 1947 by the Anglican Church. They were told they could stay there indefinitely (so long as they were practising Anglicans) and until the time the land was sold have practised subsistence on 283 hectares, using the rest to graze cattle. The sale has meant they have only 125 ha to live on. The balance, sold to Kiribati would be difficult to farm in any case, and the Solomon Islanders currently living there agree, claiming the land could support very few people. The Solomon Islanders are also in the unenviable (and insecure) position of now having to lease the land which they were permitted to keep for just 99 years, being told by the church that 'that is all they will get'.

The Anglican Church has been accused of failing in its duty to the Solomon Islanders, and of taking advantage of an unsophisticated buyer. The church claims that people could transform the landscape but appears to have no concern over the future of the Solomon Island descendants. Meanwhile, President Tong announced the formation of a committee to study what should be done with the land.[90]

The possible relocation to Fiji and the way it has been handled is ironic given Fiji's efforts to relocate many of its coastal villages across the country in the first wave of projects to protect villagers from rising sea levels.[91] The need for Fiji to deal with its own people complicates the future of the Solomon Islanders living in the area purchased by Kiribati.

It is unfortunate that people already living on the land have been so marginalized, but it is also of concern that the people of Kiribati have had such little say in their relocation, if indeed it happens. At home in Kiribati there is opposition to the purchase with people variously accusing the government of spending money on 'foolish things', and being tricked by the Church of England, which is gouging 'one of the poorest and most isolated countries in the world'. Others are even harsher, claiming the land purchase is a publicity stunt. A former president, Teburoro Tito (also past Secretary General of the PIF), said the fact that Tong has variously described the purchase as being for relocation, food security and as an investment, highlighted the fact that the government had no clear plan.

Since the 2016 appointment of a new President of the Republic of Kiribati, Te Beretitenti Taneti Maamau, there has been little media attention paid to the Fiji land purchase, and the few i-Kiribati who can get visas to migrate (e.g., just 75 per year to New Zealand) have few options.

Interestingly, in 2016, Kiribati gave assistance to Fiji of US $50,000 (approximately FJ $104,494.85) towards recovery efforts following Cyclone Winston.[92]

What might make some sense for Kiribati could be internal relocation, to North Tarawa as noted earlier, or even to Christmas Island in the Line Island group. The mayor of Christmas Island says that what Kiribati needs is more coconuts (a staple part of the diet and also an income from copra) but since people moved from Tarawa to Christmas over the past few years, demand has outstripped supply. The Mayor comments that it takes four years for a sapling to produce coconuts, so 'instead of buying land in Fiji, the government should hire people to plant trees, lots of trees'. Then, he said, people emigrating from overcrowded, flood-prone Tarawa would not only have plenty of space and drinking water, but a way to earn a living.

It seems that despite international awareness of Kiribati's possible future there is inadequate knowledge of what might be the most sensible approaches for this nation and other atoll countries such as Tuvalu and Republic of Marshall Islands. Some people are sceptical of warnings of impending doom, which can bring notoriety and fame abroad, but which also risks portraying island people as victims with no resilience or power. This is a pity as it is clear that atoll and low-lying nations, as well as all coastal areas, face a very uncertain future. There is certainly much adaptation work going on at the local level[93] and it is important this continues, within the context of peoples' own beliefs, languages and traditional knowledge.

The Future

The cases of land construction, relocation and environmental management outlined here remind us there are many adaptive approaches to facing the impending scenario of rising sea levels. Climate change is already bringing more devastating storms, inundation and consequences for agriculture and fisheries, water supply and the potential loss of land and even entire countries. Although such a future is now known (no matter that some of the visual portrayals are exaggerated), the real questions remain: should we stay or should we go? And if we stay, where and how should we live? If we go, will the future location be suitable, and what will be the impact on moving population and the hosts? Whatever the future holds for the Republic of Kiribati and others, it is important people are treated with respect, and involved in future decision-making, so they are not solely viewed as 'refugees or drowning islanders'.[94]

Some of the case studies detailed here involve technological solutions of modern, futuristic artificial islands, which will not only be extremely costly, but which appear to take little account of how people live currently, or their own knowledge of the sea and surrounding environment. Importantly they pay no heed to people's own technological solutions, their own 'adaptive strategies' also outlined, such as food preservation, artificial islands and areas of reclamations, which have existed for centuries. Additionally, recognition of current technical traditional knowledge on the state of reefs, how rapidly they grow, and can be sustainably managed, can usefully inform people that there are other options and places where people can go so that they do not have to abandon ancestral lands and surrounding sea. Instead, 'the only adaptation option of relocation'[95] is now being presented as one of the key 'solutions' for Kiribati and other countries. The purchase of Fiji land has many negative aspects and far from being the 'saviour' of Kiribati may well turn out to have implications not only for the Kiribati people themselves, but also for their hosts, unless cultural traditions can be retained thus enhancing resilience of the communities.

NOTES

1. Climate change and disaster risk reduction (DRR) given the cost of aid could be better integrated with community practice (see Bettencourt et al. 2006).
2. Garnaut (2008).
3. Morrison (2017).
4. Bryant-Tokalau (2008).
5. Donor competition has been particularly rife in the Pacific Islands with countries essentially competing for projects, and ultimately having little long-term impact on the lives of the people concerned. Commentary on the implementation, philosophy and success or otherwise of such projects has been widely made, for example, Barnett and Campbell (2010), Bryant-Tokalau (2008), M. Van Veldhuizen (2014).
6. A gyre describes the circular movement of the ocean currents. In the North Pacific the circulation of water is clockwise (Nunn 1999: 43). Chapter 'Geomorphology' in Rapoport 1999).
7. Bryant-Tokalau (2011).
8. See http://www.pina.com.fj.
9. Ilan Kelman, I. 9 August 2010 to SICRIGoogleGroups.com.

10. It is fully understood that islands do move due to their positions on tectonic plates: they are subject to rising, subsidence and outright disappearance, due to geological and climatic events, and are prone to human modification and disturbance. Such change can be both gradual and catastrophic and over hundreds of years coastal areas constantly recede and replenish (Nunn 1999: 51–53). In earlier geologic times such as in the Little Climatic Optimum of AD 750–1300, Nunn and Britton remind us of rising sea levels that will have influenced settlement patterns into nucleated groupings due to the loss of flat land (Nunn and Britton 2001).
11. Löfgren (2007).
12. Nunn (1994), p. 342.
13. D'Arcy (2006: 128). There was also, in the 1840s, a total eclipse of the sun, earthquakes and hurricanes in the Pacific region, all of which must have lead to significant community responses.
14. http:/webecoist.com/2009/04/27/12-fantastic-floating-cities-and-artificial-islands/.
15. There are cases of mangroves being reclaimed for tourism in other parts of the Pacific such as Kosrae (FSM) and Samoa but not to the extent of Fiji.
16. Interestingly there has been very little evaluation of the risk, especially of flooding that resort areas such as Denarau pose both to surrounding populations, and to the tourists and workers on the island. Bernard and Cook (2014) have commented on the risk of flooding, the cost and losses this means for tourism (the major industry) and warn of the implications across the Pacific for such a key industry.
17. http://shimz.co.jp/English/theme/dream/greenfloat.html.
18. http://lfort.wordpress.com/2010/01/21/climate-change-has-Pacific-considering-artificial-islands.
19. See Kelman (2010), Bryant-Tokalau (2011).
20. http://www.isciencetimes.com/articles/6066/20130916/pacific-island-floating-country-kiribati-global-warming.htm.
21. An artist's impression can be seen on http://news.nationalpost.com/2013/09/15/faced-with-rising-sea-levels-pacific-state-looks-to-become-worlds-first-floating-nation/.
22. http://www.isciencetimes.com/articles/6066/20130916/pacific-island-floating-country-kiribati-global-warming.htm.
23. http://www.un.org/en/ga/69/meetings/gadebate/pdf/KI_en.pdf.
24. Bryant-Tokalau (2015). 'Indigenous responses to environmental challenges: artificial islands and the challenges of relocation'. Paper presented to Pacific History Association conference, Taipei, Taiwan. November.
25. The fact that these islands could not be considered permanent raises several points in law as if not historically assimilated they do not qualify as territory, Kardol R. (1999).

26. See http://www.bbc.com/news/world-asia-38647174.
27. https://www.nytimes.com/2017/01/27/world/australia/climate-change-floating-islands.html.
28. Ivens (1930).
29. Nunn, P.D., et al. (2016b).
30. Nunn (2009a).
31. Webb and Kench (2010).
32. Pacific Islands Applied Geoscience Commission (SOPAC) 2012–SOPAC (2009) Relationship between natural disasters and poverty: a Fiji case study. SOPAC Miscellaneous Report 678. Prepared for UN International Strategy for Disaster Reduction Secretariat's 2009 Global Assessment Report on Disaster Reduction.
33. Ayres (1983).
34. Bryant-Tokalau (2011).
35. Bryant-Tokalau (2015).
36. D'Arcy (2006: 128–132).
37. Ibid.: 133.
38. Nunn (1999). Geomorphology pp. 43–55 in Rapoport (1999).
39. Hviding (1998).
40. Ivens (1930).
41. Parsonson (1966).
42. Guo Pei-Yi (2001).
43. Parsonson (1966: 5).
44. Solwara and bus people refers to those who build out over the water (sol wara) or in the bush (bus).
45. Guo (2001).
46. Hviding (1998: 259).
47. Irene Hundleby (2014: pers. comm.).
48. Parsonson (1966: 18).
49. Parsonson (1966: 18–20).
50. Walter and Hamilton are discussing as context, community-based conservation (2014: 1–10).
51. Walter and Hamilton (2014: 2).
52. See, for example, Tabe (2016), Hau'ofa (1994).
53. Bryant-Tokalau (2014b).
54. 19 March 2015. Devpolicy devpolicy@anu.edu.au.
55. MacLellan (2015b).
56. Tomlinson referred to this conflict as 'power encounters' where missionaries challenged local expectations of spiritual efficacy, by denying local sites' (such as volcanoes) original potential to evoke wonder.
57. The power and aftermath of Cyclone Winston which struck Fiji in early 2016, while still being assessed, was so devastating that it is not yet time to assess the long-term impact.

58. Campbell and Bedford (2014).
59. Campbell (2014).
60. Connell (2010).
61. Connell (2012).
62. Connell (2013).
63. Smith (2013).
64. Campbell (2014: 1).
65. Campbell (2014: 12).
66. Rakova (2017).
67. Banaba was largely ruined by phosphate mining for New Zealand and Australian farms. The Japanese occupied it during World War Two and either murdered or enslaved much of the population on Chuuk (then Truk) in Micronesia. After the war, the British who were the colonial administrator refused to return the survivors to Banaba, instead relocating them on Rabi Island which they purchased for them in Fiji. Today there are around 5000 Banabans living on Rabi with around 300 on Banaba itself. See Teaiwa, K.M. (2015).
68. Secretariat of the Pacific Community (2015). *Pocket Statistical Summary.*
69. Australian Bureau of Meteorology & CSIRO (2011: 99) Climate Change in the Pacific: Scientific Assessment and New Research. Volume 2: Country Reports.
70. Australian Bureau of Meteorology & CSIRO (2011: 100).
71. The Phoenix group of eight islands is found in the central Pacific to the east of Tarawa.
72. Tabe (2014).
73. In 1992 when I was working in Kiribati I encountered hospital syringes (needles) on a sandbar in the lagoon some kilometres from Tarawa.
74. Nunn (2009b).
75. Caritas (2014: 23).
76. Donner (2015), p. 53.
77. Webb and Kench (2010: 8).
78. Webb and Kench (2010).
79. Biribo and Woodroffe (2013) Dr Biribo in 2011 became a recipient of the 2011 Prime Minister of Australia's Pacific-Australia Award, an initiative designed to promote knowledge, education links and enduring ties with countries in the Pacific, Papua New Guinea and East Timor.
80. Options, according to Kench, would be to move to islands such as Aranuka, or to North Tarawa which is much wider and has good protection against the waves.
81. Tong (2016: 23).
82. Taberannang Korauaba (2012). *Media and the politics of climate change in Kiribati: a case study on journalism in a 'disappearing nation'* Master of Communication dissertation. Auckland University of Technology. Abstract.

83. Government of Kiribati (2013). *Second Communication under the United Nations Framework Convention on Climate Change*. Environment & Conservation Division, with assistance of Climate Change Study Team Ministry of Environment, Lands and Agricultural Development.
84. GOK (2013: 164 ff.)—see especially, section on adaptation.
85. http://www.huffingtonpost.com/2014/09/22/kiribati-president-buying-land_n_5860064.html.
86. Christopher Pala (2014). The Nation that bought a back-up property. *The Atlantic*, August 21.
87. Pala (2014).
88. Hermann, E., and Kempf, W. (2017). Climate Change and the Imagining of Migration: Emerging Discourses on Kiribati's Land Purchase in Fiji.
89. Refer Bryant-Tokalau (2014b).
90. http://www.ipsnews.net/2014/06/kiribati-president-purchases-worthless-resettlement-land-as-precaution-against-rising-sea/.
91. http://aosis.org/reports-fiji-latest-country-to-relocate-climate-refugees/.
92. Pacific Climate Change Portal (2016), https://www.pacificclimatechange.net/news/ accessed 15 March 2017.
93. Adaptation strategies in Kiribati include a number of both soft and hard responses, coordinated by the *National Adaptation Steering Committee* (NASC), a *Climate Change Study Team* (CCST) and a number of government departments and organizations. Projects on water access, treatment and monitoring, engineered seawalls, coral reef monitoring, health strategies, environmental education and mangrove replanting supplement the many reports produced by the teams and committees (GOK 2013).
94. Donner (2015: 57).
95. Nunn (2009a).

CHAPTER 4

Handling Weather Disasters: The Resilience and Adaptive Capacity of Pacific Island Communities

Abstract Handling weather disasters through the lens of earlier disasters, as well as recent ones such as Cyclone Winston in Fiji, and other recent floods, the resilience and adaptation of peoples are examined here. How people prepare for, then cope with disasters, utilizing indigenous knowledge systems is shown as still having significance in the contemporary era. Modern disaster planning needs to take account of such knowledge and especially acknowledge that people have far more awareness of how to prepare than planners often give communities credit for. The need to understand customary management practices, including land tenure systems and leadership should not be ignored.

Keywords Weather disasters • Cyclones • Floods • Traditional coping

INTRODUCTION

In February 2016, Fiji suffered from Cyclone Winston, said to be the strongest cyclone ever (at least in recorded history) to have made landfall in Fiji or indeed in the southern Pacific. The run-up to Winston was long and slow. Over several days the system gradually developed, moving southeast, reaching greater intensity by February 12, with winds of up to 175 km/h (110 mph). The system changed track a couple of times and then intensified, developing into a Category 5 system a week later

© The Author(s) 2018
J. Bryant-Tokalau, *Indigenous Pacific Approaches to Climate Change*, Palgrave Studies in Disaster Anthropology,
https://doi.org/10.1007/978-3-319-78399-4_4

(February 19 and 20). It then passed directly over the island of Vanua Balavu in the Lau group to the south of Fiji where recorded wind gusts of over 306 km/h (190 mph) created havoc.

The next day, the cyclone reached peak intensity with sustained winds of 230 km/h (145 mph) and very low pressure before making landfall on the main island of Viti Levu. Damage to settlements was extensive and there were 44 deaths. As many as 40,000 homes are thought to have been damaged or destroyed and around 350,000 people (approximately 40 per cent of Fiji's population) were affected by the cyclone. Although it is difficult to assess the cost of damage, about FJ $3 billion (US $1.4 billion) is thought to have occurred. External aid has been considerable, but for many in Fiji, particularly on the island of Koro, and on the northwest coast of Viti Levu, in the Ra province, people remain living in tents, with no water, schooling or other services.

> *It's miserable. Small babies, older people, water just floods through and they end up sleeping on some of these wet floors because there's very little option. … The area's evacuation centres—schools and halls—were destroyed during Winston and have yet to be replaced. … It has been very exhausting the whole year for them, so they're getting sort of adapted to living like this. … There's a lot more sickness. You see much more fungal growth with the wetness all the time, much more skin diseases we are seeing.* Sashi Kiran, Director of Foundation for Rural Integrated Enterprises & Development (FRIEND) (http://www.radionz.co.nz/international/pacific-news/324871/resilience-amid-the-struggle-one-year-on-from-cyclone-winston)

Such has been the picture for parts of the Pacific over the past year. Both Fiji and Vanuatu have faced violent storms. Despite the recent El Niño weather pattern that has brought lengthy periods of drought, crop failure and starvation to parts of the Pacific, floods and other disasters are also becoming more common and more destructive. Impacts of long-term climate change include not only sea level rise and fiercer storms, but also growing vulnerability to the destructive force of such floods such as infrastructural damage, growing poverty and hardship, exposure to communicable diseases and of course damage to cash and local crops and to the wider economies.[1] Pacific countries have faced numerous floods in the past decades. Most recently, Honiara in Solomon Islands and Fiji (especially the towns of Nadi and Labasa) suffered massive deluges with a loss of bridges, homes and roads, agricultural land, as well as a number of deaths and relocations, especially in urban areas.

What is causing this increasing devastation? Small, low-lying islands, coastal areas and atolls of the Pacific, are very much in the frontline of climate change, and globally the impacts are being felt everywhere. Harsher weather patterns and growing numbers of deaths and devastation are increasingly common and so the dilemma becomes one of not only responding with emergency relief, but also of building up resilience, for both communities and formal structures. Learning, sharing and applying lessons from existing knowledge and practices have to be part of flood response and preparation.

Weather-related emergencies and disasters test people's resilience. How that resilience takes shape can depend on community networks and experience developed over generations. That strength of community has certainly assisted people in adapting to the increasing number of floods affecting Fiji, but this was sorely tested with Cyclone Winston. It can be argued the poor and marginal are more vulnerable to the impacts of disaster, including death, illness and loss of homes but they are not necessarily without knowledge of what to do. A key component of people's 'adaptive strategies' for dealing with disasters includes the use of traditional knowledge and practices, even where all who practise such strategies are not indigenous. This does not necessarily mean, as in Fiji, that people are less resilient or less able to cope with the difficult conditions. Examples from rural areas, as well as informal urban settlements on Viti Levu, and more isolated islands, will, in the following chapter, illustrate the adaptability of people dealing with major environmental change.

People's adaptations to disaster should be viewed in the broader context of climate security. As noted earlier, climate change is more than rising sea levels. It includes the impacts of these changes on food supplies, the ability to grow food, access to safe water, as well as the basic necessities of life such as shelter and access to health care.[2] Resilience to disaster depends very much on robust systems, resourcefulness, rapidity and redundancy,[3] and it will be argued here that these four elements are widely present in many aspects of life in the many, diverse islands of the Pacific. Even where people are globalized, urbanized or 'disconnected' (such as with the relocation of 'climate refugees')[4,5] coupled with the loss of statehood and sovereignty, isolation from land, culture and issues of food security and resilience remains.

Climate-related disasters also place considerable strain on the budgets of nations. Humanitarian and military assistance is more frequently required to deal with such events, placing pressure on island nations' coffers, as well as stretching the capabilities of organizations providing development assistance.[6] Donors also act strategically, and aid at the time of a disaster can lead to long-term 'strategic influence'. In small PICs, this may have negative implications for future development strategies.

Fiji is no exception to the wide range of security issues due to weather-related disasters. While floods and cyclones have always been a feature of life in Fiji, the intensity and frequency appear to be increasing.[7] Although a cyclone such as Winston had been predicted as the impacts of climate change became more obvious, it is also increasingly understood that the impacts of flooding are being exacerbated by human activities.

Flooding in Fiji

Fiji often has to deal with major floods due to its location in the central south Pacific, and its vulnerability to tropical cyclones. In the 1840s several cataclysmic events hit Fiji, including in the Rewa Delta where massive floods left few unscathed.[8] In the one hundred years between 1875 and 1975 around 125 cyclones hit the wider Fiji group,[9] although these were not evenly or annually spread. Such events place severe strain on traditional community structures yet societies adjust and survive through various forms of adaptation. These can be as varied as new coastal developments and the building up of land, utilizing long-held traditional beliefs and knowledge, the forging of new alliances and migration. In Fiji all of these have occurred. Some decades have been left unscathed by weather disasters, but at other times, such as in 1972, there were several major cyclones, requiring both immediate and long-term community and international response. On average, 12 to 15 tropical cyclones affect Fiji each decade and every one causes high intensity rainfall.[10,11] Not counting Cyclone Winston in 2016, some of the most serious cyclones to strike Fiji have included the 1886 cyclone, which caused great damage to the islands of Gau, Taveuni and Mago,[12] and the 1931 flood which followed a cyclone in late February and affected the urban centres of Ba, Nausori, Sigatoka and Nadi leading to 225 deaths[13] (Fig. 4.1).

The impacts of these cyclone-induced floods are nearly always severe, with particular problems in the town of Nadi in the west, and in the northern division: few areas are left unscathed. Usually there is widespread flooding in low-lying areas throughout the country and roads are closed or difficult to use because of downed power lines and debris. Power can be

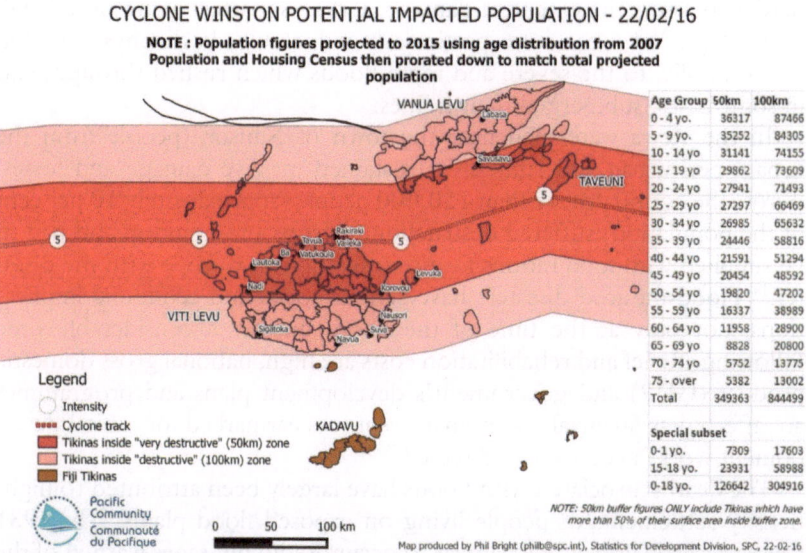

CYCLONE WINSTON POTENTIAL IMPACTED POPULATION - 22/02/16

NOTE : Population figures projected to 2015 using age distribution from 2007
Population and Housing Census then prorated down to match total projected
population

Age Group	50km	100km
0 - 4 yo.	36317	87466
5 - 9 yo.	35252	84305
10 - 14 yo	31141	74155
15 - 19 yo	29862	73609
20 - 24 yo	27941	71493
25 - 29 yo	27297	66469
30 - 34 yo	26985	65632
35 - 39 yo	24446	58816
40 - 44 yo	21591	51294
45 - 49 yo	20454	48592
50 - 54 yo	19820	47202
55 - 59 yo	16337	38989
60 - 64 yo	11958	28900
65 - 69 yo	8828	20800
70 - 74 yo	5752	13775
75 - over	5382	13002
Total	**349363**	**844499**

Special subset		
0-1 yo.	7309	17603
15-18 yo.	23931	58988
0-18 yo.	126642	304916

Legend

- ◯ Intensity
- ●●●● Cyclone track
- ▮ Tikinas inside "very destructive" (50km) zone
- ▯ Tikinas inside "destructive" (100km) zone
- ▮ Fiji Tikinas

Pacific Community
Communauté du Pacifique

0   50   100 km

NOTE: 50km buffer figures ONLY include Tikinas which have
more than 50% of their surface area inside buffer zone.

Map produced by Phil Bright (philb@spc.int), Statistics for Development Division, SPC, 22-02-16

Fig. 4.1 Cyclone Winston path. Source: Phil Bright, SPC. With permission

cut for 24 to 48 hours in urban areas (much longer in rural areas) and some bridges and roads are washed out. People are urged to relocate to emergency shelters (e.g. there were 3500 people in the Nadi shelters in December 2012) and there are always several deaths by drowning. Flights at the international airport of Nadi can be delayed (and also near Suva where Nausori airport can be flooded). Normally the Nadi River breaks its banks in several places and large areas of the approximately 43,000-person town are flooded which is further exacerbated by the high tide. Recent similar scenarios have witnessed flooding in both February and March 2013, January and December 2014 and February 2016.

In January 2009 Fiji dealt with one of its worst floods since 1931. Over 400 mm of rain fell over a 48 hour period leaving a dozen people dead and more than 10,000 displaced, with severe impacts on crops, infrastructure and health issues due to contaminated water and disease. These impacts were serious in both rural and urban areas but for those in towns, particularly in marginal areas such as informal settlements, it was much harder to assess the extent of damage. Urban squatters living on coastal, often degraded areas were particularly vulnerable, both physically and economically, to the severe and rapid floods which rushed through, and in some cases, submerged their homes.

In the Rewa watershed near the town of Nausori, people from the urban, peri-urban and rural areas sustained massive damage and losses. Across the country more than 120,000 people (approximately 10 per cent of the population) suffered destruction of crops and housing, and had to be supplied with food rations for up to six months to cope with the disaster.[14] Flood-induced disasters have serious social and economic implications, not only at the time of the flood, but for a considerable time following. Relief and rehabilitation costs are high, national gross domestic product (GDP) and government's development plans and programmes are affected as financial and human resources earmarked for capital development works need to be redirected.[15,16]

The deaths associated with floods have largely been attributed to high-density settlements of people living on exposed flood plains.[17] In 1931 there was little notice of what was to occur. Radio messages warned of the cyclone but not the flood, which came up rapidly on the Ba flood plain during the night. Many people had no radio.[18] It is also very likely the Indian settlers had no knowledge of the flood before it hit due to limited community links and the absence of radios. Fijian villages may have been more aware as the *buli* (village head) was contacted by their District Commissioner, but a number of other reasons can be attributed to the fact there were fewer deaths of ethnic Fijians. They tended to live on higher ground having lived a long time in those locations, and on hearing of the flood would have beaten the *lali* (drum) to warn surrounding villages. In addition, popular press claimed Fijians were often able to swim, as opposed to Indo-Fijians who generally could not.[19]

In Fiji, as in the wider Pacific, human settlement patterns have changed considerably over the last century, particularly in the past 50 years, with increasing urban concentration and growing numbers of displaced rural people living in informal communities (sometimes known as squatter settlements).[20,21] Most of these are located on flood plains that are often the only place settlers can get access to land. It is argued that this unprotected location of many new urban dwellers makes them highly vulnerable to major flood disasters.[22]

How People in Fiji Deal with Disaster: Traditional Ways of Coping[23]

One focus of this chapter is the way people cope or deal with climate-related disasters using Indigenous Knowledge Systems (ILK). Although there are limitations to ILK, especially through globalization, language loss, migration and pertinently, loss of respect, as long as cultural values of sharing and valuing community remain and those collaborations continue, then, as noted by McMillen et al. (2014), resilience, and thus the ability to adapt, will be maintained. It is not only rural, traditional village people who have local knowledge and ways of coping, but many urban people also retain a number of strategies that enable them to survive a major disruption to their lives.[24] A lack of attention to urban dwellers in strategizing for disasters can mean there is a tendency in current disaster and climate literature to focus narrowly on rural and 'traditional' methods utilizing traditional ecological knowledge only with the possible outcome that a major urban disaster is a very likely scenario in the future.

Disaster planning these days tends to focus on coping strategies of rural peoples in both planning for and dealing with disasters[25] (particularly where countries are very dependent upon primary industry, usually agriculture, as the mainstay of the economy). Earlier studies emphasized food security through methods of storing surpluses, preserving, protecting and preparing gardens, and always ensuring diversity of production. Others focused on inter and intra-community cooperation through practices of ceremony and exchange.[26]

More recently writings on vulnerability and disaster response have stressed levels of poverty in Pacific countries and until the last decade that focus has been in rural areas where dependency on the primary sector has been very closely associated with poverty and the impacts of disaster. In Fiji, a report by SOPAC (Pacific Islands Applied Geoscience Commission) linked low incomes in the primary sector with hazardous locations and vulnerability.[27] Although the report provided a good summary of hazards, vulnerability, policy implications and official responses, the outcomes of disasters for the urban poor and attention to peoples' traditional knowledge was not a focus.[28,29] Now, through an examination of resilience, it is understood that people are more knowledgeable about their changing environments than previously known, and that by examining wider aspects of peoples' lives, particularly through broader 'knowledge-practice-belief systems' applied, for example, in managing local biodiversity, adaptation to climate change disasters can also be better understood.[30]

In this chapter, case studies focusing on community cooperation and practice will be presented. Descriptions of community responses to floods and cyclones not only provide lessons that may assist in planning disaster responses today, but also give insights into the long-term resilience of communities, frequently based on indigenous knowledge and practice.

Cyclone Val 1975

How people respond to disasters, particularly cyclones, is crucial to understanding how appropriate responses can be prepared. In the mid-1970s, while working on the MAB Hurricane Hazard Project,[31] Muriel Brookfield produced an insightful report on how the people of Lakeba, Fiji responded to Cagi Laba (Cyclone Val) and its aftermath.[32] In that paper, based on interviews with the people of Lakeba, she detailed a range of responses to the cyclone. These included fight and flight responses such as migrating away due to fear, making plans for the future and also a fatalistic acceptance of the will of God. Measures undertaken by the people themselves before, during and following the cyclone were outlined in some detail, providing a useful snapshot more than 30 years later when agencies continue to struggle to reach people and to provide the most appropriate assistance. Tellingly, Brookfield also assessed what turned out to be ineffectual assistance to the people of Lakeba, with limited information, highlighting the sense of abandonment and isolation felt by the people of Lau.

Once the time is right for an analysis of Cyclone Winston it may be possible to compare what happened with the human responses to Cyclone Val. There can be seen now though some commonalities and types of human endurance and beliefs that can be discussed. The 1975 cyclone and the aftermath demonstrated the problems of isolation and lack of communication faced by peoples in the outer islands, especially during a period of disaster. It also highlighted the fact that people, in the wake of disasters, and although resilient in many ways, have to fend for themselves for more aspects of daily life than perhaps was the case in the past. Brookfield's Lakeba study also demonstrated the isolation felt by outer islands. The island had not endured a cyclone for 27 years at the time of cyclone Val[33] and had in many ways, been lulled into a sense of false security. It was a time in Fiji's history when all seemed to be going well. Copra prices were high, the village was home of the Tui Nayau (Ratu Sir Kamisese Mara, also prime minister),[34] and perhaps there was almost a feeling of protection, even immunity to danger. But despite this, communication systems were erratic and the people were not informed of the change in track of the cyclone until it was almost too late to do much to prepare for damage. People could soon see for themselves not long before the cyclone hit that the damage was likely to be severe and they did what they could by putting up shutters or boards on houses and reducing the height of plants such as cassava to almost ground level (so they would not be completely uprooted and destroyed). People also moved to stronger buildings where they could cut large branches from threatening trees, bring in food supplies, water and cooking pots to the houses, and usher children home from school where possible. The stories written after the cyclone by the children of Tubou in Lakeba at Brookfield's encouragement provide very graphic and immediate descriptions of what occurred, and gave the children an outlet for what they had experienced. Although there were no deaths on Lakeba, there was large scale destruction with the loss of livestock, homes, boats and food gardens. Brookfield and others also reported several stories of great courage, such as the woman who, in the last stages of a difficult labour, walked uphill through the storm to the hospital where she safely gave birth, while her husband, the headmaster of the boarding school looked after his 58 charges back in the school.[35]

There was extensive damage in the Lau group with more than 300 houses destroyed, 87 of them on Lakeba. School classrooms in Tubou village were severely damaged, but away from the village there was even greater impact on gardens and cash crops.[36] In Lau, people were very dependent for their income on copra and yaqona and both were significantly damaged. But it was the food crops that were the worst affected, and although root crops could be harvested for a few weeks before they rotted in the ground, after that there were few supplies available. Challenges to food security were made greater for a time as fish stocks were also badly affected with nets and boats destroyed. EMSEC (the National Emergency Service Committee) had great difficulty in reaching many parts of the Lau group, needing to assist firstly those most badly hit such as on the island of Kabara (also in Lau), so essentially people were left alone to cope as best they could. Self-help was encouraged,[37] and indeed people did re-build their houses and replant their crops while waiting for emergency aid, but their resilience was sorely tested. Some were so badly affected by the cyclone they suffered nightmares and lost interest in their surroundings and others even considered moving away either for children's schooling, or for wage earning jobs, for as they quickly found, they had no source of income to pay for other necessities such as fuel, food[38] and children's school supplies.[39] Resentment was also a feature of the response for many felt that they were expected to work in order to 'earn food' while waiting for life to get back to normal.[40] Although people understood, and indeed did work hard in the ensuing months the long hours meant they were less able to work formally for an income, much needed for supplies and school fees.

The long-term impact of cyclone Val on Lakeba was significant. People did leave, many of them to return only infrequently, but there were also changes to the community such as new and stronger building styles, such as the construction of one very strong hall to house all villagers, more care for the safety of livestock and planting of less vulnerable crops. There was also an enhanced dependency on imported and store-bought food, and the perceived need and desire for a formal monetary income.[41] As happens everywhere with such dramatic change, social change came about perhaps more rapidly in the aftermath of the cyclone, with the 'unemployed, ill-educated, bored youth with dreams of city-wealth and living… breaking away from the old ways'.[42]

But it was not only the material changes brought about that were of long-term significance. There appeared to be, according to Brookfield, greater credit being given to traditional ecological knowledge as predictors. Although Brookfield did not refer to these signs as TEK, they would today be viewed as a resurgence of the knowledge that although always there, had possibly been forgotten, or at least not treated seriously on an island that was in many ways privileged from the long absence of disasters, until Cyclone Val arrived. It was noted that people were more alert to the likelihood of cyclones recurring and then more openly discussed potential signs as, such as very hot and wet weather, the high abundance of mangoes and breadfruit, an increase in land crabs and changes in the flight paths of sea birds.[43] Because of the long period of complacency on Lakeba, few had been taking note of such signs before the 'Cagi Laba', but people became more aware afterwards, although some still regarded these occurrences as myths.[44]

A resurgence of interest and attention to local changes in the environment could be viewed as reminiscent of earlier, pre-colonial, pre-aid agency times where people were most likely more in tune with their surroundings, with a better understanding of the impact of both human and natural disasters on local resources. The long neglect of such signs may also have brought about complacency and possibly a false sense of security in the community perhaps dulling their resilience and preparation.

Brookfield's account of Cyclone Val's social and physical impacts serves to highlight the difficulties of isolated people in the face of disasters that, in the modern era of centralized disaster relief, military assistance and the global reach of the media, are mostly dealt with centrally but which tend to neglect those on the periphery. It also shows how people, in the absence of assistance and even communications, have to be self-reliant. Brookfield partially attributes the difficulties faced by Lakeba to the inherited centralized colonial structure,[45] and highlights the importance of appropriate, speedy aid and a focus on more environmental awareness (and thus, one assumes, recognition of how damage can be more severe where environments are not protected).

As with the impacts of Cyclone Winston on islands such as Koro, the Lakeba case provides a base for discussion of how the poor, isolated and the marginal deal with a disaster in circumstances where they have no real access to central government and emergency services, few connections with people who can assist, and frequently few personal resources that can be utilized during and after a major disaster. The people of Lakeba were not all wealthy, despite their connections with the Tui Nayau. Although they had had high incomes from copra in the years preceding Cyclone Val, many were rendered very poor afterwards, facing dietary shifts on top of lower wages and failed crops.[46] They were isolated and far from emergency assistance. Resilience, self-help and traditional ways of coping were available to them, but overall the community still suffered, both psychologically and physically.

The case of Cyclone Val and Lakeba is useful for understanding the impacts of modern disasters in Fiji but the work is barely remembered today.[47] Now, when planning for disasters, much of the emphasis is on relocation, the likely impacts of cyclones on tourism and loss of revenue. Attention to people's knowledge and experience of severe weather events is often overlooked, especially in urban areas. Traditional knowledge systems, although often mentioned as something to be preserved and replicated, are sometimes not transferred to modern living conditions, or indeed to current academic literature and the media. It may be time to resurrect a 'combining of disaster research' with 'participatory action',[48] or at the very least, to give recognition that there is more knowledge of dealing with disasters than planners give people credit for.

An excellent basis for this recognition could be by paying more attention to Pacific Island insights and knowledge of customary resource use and their resilience to disasters, including those related to climate change. There remain researchers and donors who prefer to view Pacific futures as lacking in hope and possibilities for dealing with the impacts of climate limited[49] but in contrast, a recent paper in *Ecology & Society* produced from a workshop in Fiji on local ecological knowledge, it is asserted that in order to understand socio-ecological systems and how they can be resilient to climate change, it is crucial to recognize the interrelatedness of many areas, including local expertise, customary resource management and practice, and the roles of leaders and institutions. Such recognition of people's existing knowledge systems (including of seasonal cycles, ecological processes etc.) are relevant to understanding resilience and adaptation to climate change.[50] Because societies are generally small, the authors argue that local people are more in tune with changes and threats. Adaptation, therefore, does happen over time. The examples given in the paper include historic voyaging, trading and exchange, adapting to marginal landscapes and maintaining wide geographic and cultural links. Systems of land tenure that included terrestrial-riparian-marine 'ridge to reef' systems demonstrate understanding of the interconnectedness of land management.[51] It is further argued that recognition of this deep knowledge held by the peoples of the Pacific can be used in current adaptation projects to climate change. McMillen, writing with several indigenous scholars including Veitayaki, Apis-Overhoff and Rupeni, each of them long-term writers and practitioners of local and traditional knowledge, provide in a 2014 paper on customary resource use and resilience, valuable detail of several forms of resilience, understanding and adaptation; customary seasonal calendars, the revival of traditional record keeping practices and the success of the Fiji Locally Managed Marine Area (LMMA) Network[52] to name a few.[53]

Leadership, both customary and modern, is another key area and the authors are honest in their assessment that weakened leadership is a challenge these days and that sometimes chiefs do not garner as much respect as they did previously.[54] But alternatives (such as the churches, which can mobilize large group activities), new leaders (not necessarily chiefs) and community groups, which include women and youth, are providing continuation. The writers do recognize a number of limitations in these Indigenous Knowledge Systems (ILK), especially through globalization, language loss, migration and pertinently, loss of intergenerational respect, but so long as cultural values of sharing and community connectedness remain and the collaborations continue, then they feel that resilience, and thus the ability to adapt, will be maintained[55] and hopefully become stronger. Veitayaki, in his writings and practice has constantly emphasized the growing strength and importance of recognizing and practising TEK and ILK in local approaches to climate change and resource management. Through his own continuing engagement with his home community of Gau in Fiji, Veitayaki demonstrates very well the adaptive capacity of communities in not only handling weather disasters, but also the possibilities for managing coastal resources where formal, institutional structures frequently do not have such capacity.[56] His work demonstrates that along with the creation of new forms of artificial islands, adaptation and the building up of land and increased planting of mangroves throughout the Pacific, in the contemporary context as well as historically, practices involving less dependence on imported and inappropriate practices such as artificial coastal seawalls can be effective in preparing communities for coastal inundation.

NOTES

1. Bell et al. (2016) warn also of potential destruction of the newly revived cacao production in places such as Vanuatu and Fiji.
2. For example, see Morrow and Bowen (2014).
3. Ikeda (2014), page 43.
4. Connell (2010).
5. Wyett (2013).
6. Pacific Institute of Public Policy (2012). 'Climate Security: a holistic approach to climate change, security and development' Discussion Paper 23, Port Vila, page 3.
7. Gero, Fletcher, Rumsey, Thiessen, Kuruppu, Buchan, Daly and Willetts (2015).
8. D'Arcy (2006: 128).
9. McLean (1977).
10. Raj (2004).
11. Yeo and Blong (2010), page 657.
12. McLean (1977), page 17.
13. In the past forty years since independence many other severe floods have affected Fiji with much damage and loss of life. Between 1983 and 2003 there were seven major events including the Tropical Cyclone Kina flood (January 1993), that followed a severe drought induced by the El Niño event of 1992/93. This is considered to be the most severe in recent history, with damage amounting to some US $100 million and 23 deaths (Bryant-Tokalau and Campbell 2014; Raj 2004: 2).

14. Bryant-Tokalau and Campbell (2014).
15. Raj (2004).
16. Bryant-Tokalau and Campbell (2014).
17. Yeo and Blong (2010: 661).
18. Yeo and Blong (2010: 668).
19. Yeo and Blong (2010: 669).
20. Chandra (1990).
21. Haberkorn (2008).
22. See, for example, the paper by Yeo and Blong (2010).
23. The failure to recognize that people in urban areas also have indigenous knowledge is a key oversight in disaster management (Bryant-Tokalau 2016c).
24. The issue of what is 'urban' and how this leads to differing responses is also to be examined here.
25. Writers such as Timothy Peter Bayliss-Smith (1977), Hurricane Val in north Lakeba: the view from 1975, pp. 65–98 in McLean, Bayliss-Smith, Brookfield and Campbell (1977), Campbell (2006) and organizations such as SOPAC (2009).
26. Ravuvu (1988).
27. SOPAC (2009: 39).
28. Bryant-Tokalau and Campbell (2014).
29. Campbell (2006) notes in his papers on disaster preparedness the most important issue of food security and how systems of surpluses, preservation, the fragmentation of gardening land and diversity of production all contribute to better resilience in the face of disaster. He also discusses inter and intra-community cooperation through traditional ceremonies and cultures of exchange, as well as ways of predicting future storms through observations of indicators such as sumptuous fruiting of trees and changed behaviour of fish and birdlife. Historical building practices developed over centuries in response to the environment and particularly to withstand adverse weather conditions are important in Campbell's work where he describes features such as the lashing of roofs and deep house posts indicating responses to fierce weather.
30. McMillen, Ticktin, Friedlander, Jupiter, Thaman, Campbell, Veitayaki, Giambelluca, Nihmei, Rupeni, Apis-Overhoff, Aalbersberg and Orcherton (2014).
31. Brookfield (1977).
32. M. Brookfield (1977), and see Bryant-Tokalau and Campbell (2014) in their paper on coping with floods.
33. Refer Bayliss-Smith, same volume (1977: 70–72).
34. Muriel Brookfield (1977: 103).
35. Brookfield (1977: 107–110), Bryant-Tokalau and Campbell (2014: 140–141).

36. Bayliss-Smith (1977: 72).
37. Brookfield (1977: 120).
38. Bayliss-Smith (1977: 80).
39. Brookfield (1977: 120–123).
40. Brookfield (1977:123), Bryant-Tokalau and Campbell (2014).
41. Brookfield (1977: 125), Bryant-Tokalau and Campbell (2014).
42. Brookfield (1977: 139).
43. Brookfield (1977: 134).
44. Bryant-Tokalau and Campbell (2014).
45. Brookfield (1977: 143).
46. Bayliss-Smith (1977: 82–91).
47. Bryant-Tokalau and Campbell (2014: 142).
48. Kelman, Lewis, Gaillard and Mercer (2011).
49. Bell et al., for example, in 2016 presented a relentlessly negative view of the future of PICs in the face of climate change, but this is not borne out by what people have done and continue to do for themselves, as well as the views of indigenous researchers.
50. McMillen et al. (2014: 44).
51. McMillen et al. (2014: 45).
52. Veitayaki (1997: 6).
53. Fiji's LMMA areas cover around 800 of Fiji's approximately 10,000 villages, but looking at locally managed coastal areas, around 75 per cent are covered, a far higher success than elsewhere in the Pacific (Govan, Pers. Comm. April 2017).
54. There is some discussion in Fiji in recent times about removing the role of chiefs in the granting of fishing licences. This could be extremely damaging as communities will then lose their rights to manage coastal areas (Govan, Pers. Comm. April 2017).
55. I would further contend that the failure (by omission in planning) to recognize that people in urban areas also have indigenous knowledge is a key oversight in disaster management.
56. See, for example, Veitayaki's paper on mangrove management, based on his work in Gau and on Viti Levu in Fiji. This has resonance for the wider Pacific (Veitayaki et al. 2017).

CHAPTER 5

Indigenous Knowledge Systems and Urbanization: Relocation, Planning and Modern Disasters

Abstract The focus in this chapter is on urban populations so frequently underserved, but also underestimated, in their ability to respond to major disasters. As the Pacific becomes increasingly, and often more informally, urbanized, urban dwellers can be overlooked in disaster planning, resulting in large loss of human life and property. But so too do urban dwellers have strong social networks and they find ways of coping with community and institutional cooperation.

Keywords Urbanization • Informal settlements • Indigenous knowledge • Traditional networks

In the 21st century it is clear that in the Pacific, as elsewhere, the synergy of 'natural disasters, rapid urbanisation, water scarcity, and climate change have emerged as a serious challenge for policy and planning'.[1]

There is a great deal of emphasis in academic and planning literature on the rapid and growing urbanization of the Pacific.[2,3] Urbanization is viewed as a 'problem' meaning that people living in the growing urban areas are less able to 'cope' with the rapid pace of globalization, and most certainly will struggle to adapt to the impacts of climate change.[4] Indeed, PICs are quite heavily urbanized with many having more than 50 per cent of the population living in towns, or rapidly growing in the case of

© The Author(s) 2018
J. Bryant-Tokalau, *Indigenous Pacific Approaches to Climate Change*, Palgrave Studies in Disaster Anthropology,
https://doi.org/10.1007/978-3-319-78399-4_5

Melanesian countries. Whether or not people can cope and maintain their resilience and capabilities for adaptation in the face of climate change should be carefully examined.[5]

Pacific towns have changed a great deal in the past few decades. People have increasingly become urban for many reasons including education, employment opportunities, and significantly, because of agricultural and land changes. In Fiji, for example, the expiry of cane leases[6] has forced generations of largely Indo-Fijian farmers to become urban dwellers. There is also throughout the Pacific the age-old global desire of people to migrate in the hope that life becomes better. Once in town, many are unable to afford to purchase or even rent land and formal housing and often end up living in informal settlements in marginal areas, prone to the impacts of climate-induced disasters. Honiara in Solomon Islands and Nadi, Ba and Rakiraki in Fiji are just a few towns that have faced severe flooding, with deaths and loss of livelihoods in the past two to three years.[7]

In this chapter, implications for future disaster planning are outlined with particular emphasis on the growing urban populations of Fiji,[8] where people, far from lacking in knowledge of how to deal with weather-related events, are generally able to cope (unless of course the disaster is of an unprecedented magnitude, such as Cyclone Winston in 2016).

Urban dwellers in the Pacific often know more about their local environments than sometimes imagined. Very few people helplessly wait for assistance in the event of a disaster while others may even welcome it as a 'God's will'.[9] Some urban-based literature also assumes that informal settlers have limited social networks and lack cohesion.[10] New migrants are often believed to have less understanding of local conditions, compounding the many inequalities faced by those on the margin but this is in many ways an external response to urban planning in the Pacific. Pacific urban dwellers do not necessarily fit the stereotypes portrayed of other 'indigenous' communities where they are not the dominant group. They are not 'fourth world' and nor are they marginalized in terms of culture and language. Urban areas may have once been 'colonial constructs' but since independence they are rapidly becoming 'local' or indigenous. In most Pacific countries, governance at all levels is local and the everyday language is the dominant language of the country. In Fiji urban areas, traditional networks and concepts of sharing such as kerekere have broadened to become more inclusive of different populations, taking on meanings wider than of kinship relationships.[11] Along with those wider community connections, NGOs are increasingly home-grown, and tend to work closely with communities, both urban and rural, offering alternatives and building upon what the people can, and do undertake for themselves.[12] What is happening in Pacific urban areas is moving further away from the type of western 'development aggression' portrayed by Allen[13] when talking of disaster planning.[14]

It is most often the informal ('squatter') settlements that are affected by flooding and other climate-related events, leading to major implications for the poor and more vulnerable people who are thought to have less resilience and fewer ways of adapting to such sudden disasters. Implications for the poor are multiple. Following severe storms, ongoing unemployment, loss of jobs and crops, affecting not only the nation's economy but also personal incomes can be severe. So too are difficulties of access to education and health services, as well as the inability to afford new costs and stresses associated with flood losses. As urban fringe populations are rapidly growing, they challenge the provision of water supply and quality and other infrastructure. Housing is not only informal, but also has insecure tenure making tenants vulnerable before and after disasters.

Severe flooding in Honiara, Solomon Islands, over the past few years brought a devastating toll. In 2014 Cyclone Ita led to dramatic flooding especially among the rapidly growing squatter settlements in an urban area where land is scarce and rents are too high for migrants.[15] Around 30 people were killed as a result of the floods, 52,000 people in the town area were affected and 4000 ended up in temporary shelters. In Honiara, capital of Solomon Islands, difficulties in accessing leases due to complex land claims has meant a long slow process of rehabilitation.

Extreme Floods in Pacific Towns

Although this book has emphasized the role of traditional knowledge as fundamental to adaptation to climate change in the Pacific, it is difficult to find written evidence of traditional responses in urban areas. There are many traditional coping mechanisms, such as those outlined in Chap. 4, when faced with a super cyclone like Winston, where even the settlers in Fiji towns found it difficult to cope. The Republic of Fiji is still (in 2017) facing many months of rehabilitation following the cyclone. The complexity of the three hundred islands, with several of the most populated bearing the brunt of the cyclone, brought a long-term challenge with lengthy and frustrating periods of waiting for rehabilitation. Of the population of Fiji (around 900,000), 350,000 were living in the path of the cyclone and very few would have been completely unaffected. There were 42 deaths confirmed, around 130 injuries, up to 62,000 people living in 900 evacuation centres, and schools, crops and homes destroyed with damage estimated at around FJ $1 billion[16] There was 100 per cent crop damage in some areas, meaning that many staple crops and imports had to be provided for some months. At least 87,000 households were in need of relief support, and some, especially on remote islands, and in the province of Ra that includes the town of Rakiraki, still need assistance, especially housing.

Winston was simply so powerful that no matter where people lived, whether in the interior or on the coast, it cut a wide and destructive path through the island group, completely devastating some areas. Inevitably there were some critics of government and individual preparedness and of subsequent humanitarian assistance, but it is hard to imagine how people and their structures could withstand the full force of such winds. People did survive though and the death toll was relatively low. Other impacts, such as on health (both physical and mental), food security and water, will last for some time and should give pause for reflection on how communities can deal with such disasters.

Why urban settlements are increasingly more affected by floods than rural settlements following major cyclones should be a matter of great concern. An increasing number of deaths can be attributed to flooding, but there is also an increase in communicable diseases due to overcrowding and a failure in the water supply, and thousands of people are displaced from their homes and businesses. The need for food, clean water, medical supplies and dry clothes, bedding and tents is always urgent. Small businesses (both formal and informal) in the towns could take six months to a year to recover from the floods, and may in fact never reopen. Informal enterprises such as women's market stalls and sales of handicrafts, common in urban areas, are unable to operate for months following such devastation, with major implications for social wellbeing, education of children and household survival.

The crises which are increasingly facing Pacific towns need to be examined in terms of the urban future of many Pacific countries. Much of the flood damage occurs because of the growing number of people living in difficult circumstances close to towns, and the inability of town governments and central governments, with their complex and fragmented institutional arrangements, to cope with the growing demand for access to affordable land. Many of the fast-growing towns (often situated on river banks) are also impacted by rapid deforestation upstream and the abandonment of sustainable farming practices like contour farming, 'resulting in increased erosion and siltation of water bodies'[17]. In the face of unplanned urban development government agencies find it difficult to deal with complexities of land tenure and informal developments. An increase in watershed degradation is an inevitable consequence contributing to the excessive flooding of recent years.

TRADITIONAL RESPONSES TO URBAN DISASTER

Recognition of indigenous knowledge systems is fundamental to disaster management, including in urban areas where people (no matter what their 'indigeneity') are developing new networks, indigenizing the towns and working as communities. The networks that become obvious at times of major disaster (but which are always there) are underpinned by TEK that utilizes locally referenced environmental knowledge with its history of past environmental disasters and adaptation. TEK is experiential and draws upon lessons from the past.

Despite the high levels of devastation that occur in urban areas during disasters, it is not true that there are no local, traditional or community responses. Across the Pacific communities, although highly diverse, continue to operate through kinship, and other traditional ties, and collective work continues.[18] When people become urbanized, new communities develop with new forms of engagement.[19] The challenges of urban life do not necessarily mean that family and kin connections are lost, and in many places peoples of all 'backgrounds, ethnicities and classes remain connected with kin and at the same time are developing new and powerful networks'.[20] It is also clear that in informal urban settlements such as in Suva, Nadi, Honiara and Vila for example, there remain many examples of 'traditional ways of coping', including not only kinship networks but also the development of new communities and associated sets of linkages through marriage and other relationships, whether religious or community based. Levels of resilience do exist, just as one would expect to find in traditional village communities.[21]

Traditional networks and relationships undoubtedly undergo change in urban areas, but they do not disappear, and what is strongly important to the way the settlements develop is relationships between families. In Fiji for example, most indigenous Fijians will have a *tauvu* relationship,[22] which enables sharing, and in an urban context these links may continue to be significant and widened as local knowledge is shared with new migrants.[23] Even though migrants from across the country move into the informal settlements as well as villages not their own by descent, some have strong familial relationships, but in many, with mixed membership (both ethnically and from different areas) there are established settlement committees with wide group membership. Such committees have most recently been encouraged by NGOs, but often they simply develop as people live together over a period of time. They are replicated in other urban settlements.

There are avoidance mechanisms as well. For those who have very little, ignoring traditional obligations can be a survival mechanism, but generally people do not remain completely isolated. Where communities have faced physical destruction and lives have been lost and disrupted, there are many new and local NGOs being developed to support communities. Organizations such as FRIEND in Fiji,[24] for example, with local community training and support successfully work with communities by building upon and extending existing community ties.

So too do people continue to use practices that are so normal they may neglect to see these in modern urban settings as 'traditional practices'. In Vanuatu, cultural continuity, even where cash economies place pressure on self-reliance and TEK, can still be found, as everywhere in the Pacific.[25] Knowledge such as recognizing the early warning signs of extreme weather events, implementing gardening practices such as cutting the tops of cassava plants, collecting water, tying down roofs, building houses on higher ground and moving to strong structures, for example community halls, are widely practiced. Discussions of warning signs of impending tropical cyclones are common, for example, seasonal changes in the fruiting and flowering of plants and changing behaviour of seabirds. Thaman's[26] work on urban gardens demonstrates that while many urban dwellers may not feel that they are prepared for or warned of impending disaster, many will discuss and act upon such signs with a view towards maintaining food security.

As stated earlier, there is a lot more knowledge among urban populations about how to deal with disasters than planners give them credit for and there is now a need for action—a combination of 'disaster research' and 'participatory action',[27] involving planners of disaster relief in the urban context. Most important is to incorporate traditional practices as an integral part of urban planning and disaster response.

Where formal government is including urban development into climate change policy, there is also a growing recognition of good environmental management. In Fiji for example, in written documents, 'planning and development for housing [is] to be undertaken in consideration of important environmental issues and effects of climate change'.[28] But such policy does not always mean strict adherence in practice. Public education and community practice are far more likely to have the greatest impact.

For example, with the support of the Secretariat of the Pacific Community (SPC) with New Zealand's National Institute for Water and Atmospheric Research (NIWA) a series of workshops on climate and disaster resilience in urban development planning have been held across the Pacific for representatives of national governments and Town Councils.[29] The training recognized the extreme risk faced by urban areas and was intended to ensure that future developments are resilient to climate and disaster risks. It has introduced urban planners to a range of tools developed for the Pacific Islands region, and used Nadi town as a demonstration site to support risk-informed decision-making in the approval of new developments or renovations to existing buildings. How far this training has extended to incorporate community and traditional knowledge is unclear, but this is clearly needed if deaths in urban areas during major disasters are to be avoided.[30]

With much emphasis on the rural economy and the relatively recent relocation of people to urban areas in the Pacific, many urban commentators have tended to assume that people living in towns have less resilience and lack appropriate social networks.[31] In writings on the urban Pacific the emphasis was more on negative aspects of town life and the growing numbers of urban poor, while neglecting peoples' own knowledge and resilience.[32] There was often a failure to see 'community' in urban living and people were often portrayed as poor urbanites and victims. This view was replicated at that time in preparedness planning, or the lack of it. What is better understood now however is that the majority of people never simply wait for assistance after a disaster. Although they do suffer (often disproportionately because of difficult conditions), and the communities are changed, urban people are not helpless, even in informal settlements and even when poor.

It has been widely predicted that despite all the knowledge and coping strategies developed by urban dwellers, planners and communities, the next major flooding disaster in the Pacific is most likely to occur in urban areas.[33,34] Even if these settlements are no longer settlements as previously designed by colonial powers, they are very much a product of indigenous settlement, and the predicted disasters will most likely occur, not because of a lack of knowledge, preparation and coping strategies on the part of the settlers, but because the physical event will be so powerful in the urban context. People across the Pacific, both urban and rural, are taking extreme events more seriously, in the event of a cyclone and subsequent flooding, by not simply waiting for assistance. In non-climate-related disaster events, such as tsunami warnings, people are far more likely than previously years to move to higher ground upon receiving warnings.

Ways of dealing with the threats and aftermath of disasters and a strengthening of resilience that has always been in communities are key to successful disaster response. This is not simply about coordinated institutional approaches but includes community cooperation and participation in decision-making. It is also about recognizing and promoting the types of resilience discussed earlier—having robust systems in place, recognizing and utilizing people's resourcefulness, acting rapidly and not waiting for aid, and trying not to replicate assistance, by building on what is already there.

Adaptation to climate change related disasters in the Pacific could almost be viewed as a fact of life, but urbanization (no matter its form) and globalization are challenging people's resilience. By recognizing what already exists and making it part of everyday knowledge through media and education, by understanding the interrelatedness of many areas, such as local expertise, customary resource management, knowledge and practice, as well as the roles of leaders and institutions, the 'knowledge-practice-belief systems' can be used to inform adaptation to climate change disasters.

NOTES

1. The World Bank (2016).
2. Bryant-Tokalau (2012).
3. Connell and Lea (2002).
4. See for example, Mano Mohanty (2006). Online www.hss.adelaide.edu.au/socialsciences/igu/documents/manoranjan_mohanty.
5. The issue of indigenous urbanization and what it means to the Pacific is being further explored elsewhere by Bryant-Tokalau), but it is certainly time to question what is meant by 'urbanization' in the Pacific context. The time to cast off the colonial city mantra may well be upon us with closer examination of the indigenization of the Pacific city long overdue.

6. Approximately 13,000 sugar cane leases will expire between 1999 and 2028 (Donovan Storey 2005: 15), possibly displacing as many as 75,000 people (Bryant-Tokalau 2010: 12).
7. In Fiji for example, the numbers living in the areas alongside river courses and coasts is currently unclear.
8. The urban population of Fiji is currently estimated to be 51 per cent or around 500,000 of the total population of just under 1 million people.
9. Nolet (2016) 'Are you prepared?' Representations and management of floods in Lomanikoro, Rewa (Fiji)'. *Disasters*. Link: http://onlinelibrary. wiley.com/doi/10.1111/disa.12175/full. Accessed 14 January 2016.
10. Lingam, D. (2005). *The squatter situation in Fiji*, A Report by the Ministry of Local Government, Housing, Squatter Settlement and Environment, Suva, Fiji.
11. Bryant-Tokalau, in progress. 'Multi-ethnic Suva as a Melanesian City: memories and connections'.
12. Why urban settlements in the Pacific have been more affected by floods than rural areas is a matter of great concern. The growing number of people living in difficult circumstances close to towns, rapid deforestation upstream and abandonment of sustainable farming practices like contour farming, have resulted in increased erosion and siltation of water bodies (Chandra and Dalton 2011: 8). In addition, complex and fragmented institutional land arrangements can impede flood risk reduction efforts. Often government agencies are working under several institutional regimes and the lack of harmonization can lead to confusion and lack of enforcement meaning an increase in watershed degradation and contributing to recent excessive flooding.
13. Allen (2006).
14. Allen notes the ways that root causes of vulnerability are addressed, for example poverty, and 'development aggression' (p. 82). She supports Heijmans contention that disaster response agencies increasingly used *"vulnerability" to analyse processes that lead to disasters and to identify responses'*, but at the same time, *'agencies use the concept in the way that best fits their practice—... focusing on physical and economic vulnerability'* (Annalies Heijmans 2004: 115). She agrees that social and particularly political aspects of vulnerability need to be addressed in order to make a lasting impact on overall vulnerability to disaster but warns of the danger that community-based initiatives may place greater responsibility on the shoulders of local people without necessarily proportionately increasing their capacity to formulate initiatives according to community understandings and priorities (Allen 2006: 96).
15. See Keen and McNeill (2016) which comments on the growing impacts of extreme climatic events on urban settlers.

16. UN OCHA, Situation Report 8, 26 February 2016.
17. Chandra and Dalton (2011).
18. See, for example, Bryant-Tokalau and Campbell (2014).
19. Bryant-Tokalau (2013).
20. Bryant-Tokalau (2016). http://www.mei.edu/content/map/community-responses-floods-fiji-lessons-learned.
21. See Bryant-Tokalau and Campbell (2014).
22. Tauvu refers to indigenous Fijians where people have reciprocal and joking rights with one another in certain parts of Fiji.
23. For example, see Bryant-Tokalau and Campbell, J. (2014).
24. FRIEND, Foundation for Rural Integrated Enterprises & Development is a local community development NGO based in Lautoka.
25. See, for example, the work of Granderson 2017, who, while writing of outer islands, recognized both the urban migration and the pressures placed on TEK.
26. R. Thaman in 1995 wrote of urban food gardens across the Pacific, especially Tonga and Fiji, as demonstrating continuing food security despite the rapid urbanization of Pacific populations. Later work, especially by students of the University of the South Pacific Geography department, demonstrates that the trend continues.
27. Kelman et al. (2011: 59).
28. Government of Fiji (2012).
29. The training is facilitated with support from the Asian Development Bank (ADB) and the Japan Fund for Poverty Reduction.
30. Yeo and Blong.
31. See, for example, Connell and Lea (2002).
32. For example, Bryant (1993).
33. Bryant-Tokalau and Campbell (2014).
34. Yeo and Blong (2010).

CHAPTER 6

Conclusion: What Can Pacific Island Countries Teach Others About Climate Change?

Abstract How the Pacific Islands respond to and deal with growing perceived threats of climate change can provide the wider world with valuable lessons. Incorporating traditional ecological knowledge alongside contemporary science, challenging existing formal institutions and finding new ways of dealing with a future that seems uncertain and dictated by donors and larger powers is crucial to the survival and independence of Pacific Island nations. Much of this depends on intergenerational thinking and the willingness to both fight to be heard and work with others, all fundamental to a more secure future.

Keywords Learning • Teaching • Adaptation • Indigenous knowledge • Blue economies • Carbon sink

Climate change adaptation does not have a one-size-fits-all solution. A combination of measures will be needed to ensure that there are sustained short- and long-term strategies and practices. As demonstrated in this book, what we have been observing among the island Pacific and the near neighbours of Aotearoa New Zealand are the results of long-standing observation, planning and action. While all societies are underpinned by traditional ecological knowledge that draws upon locally referenced knowledge with its history of past environmental disasters and adaptation,

© The Author(s) 2018   85
J. Bryant-Tokalau, *Indigenous Pacific Approaches to Climate Change*, Palgrave Studies in Disaster Anthropology,
https://doi.org/10.1007/978-3-319-78399-4_6

it is clear from our two books that the approaches to climate change preparedness are quite different. In the island Pacific, with its diverse cultures and multiple approaches to the impacts of climate change, there is ingrained acceptance of traditional knowledge even while the societies are undergoing upheaval in the wake of inevitable environmental and social change. There is not only strong recognition that such knowledge must be maintained, but there are also many lessons to be shared.

In Aotearoa New Zealand it is true that the country has its own set of tradition-based ecological knowledge—MEK—that references not only the environment but also the economic, social and cultural variables. This provides a platform from which to develop intergenerational adaptation strategies and practices that also draw on the experience of New Zealand's Pacific neighbours, but the entire nation needs to come on board if New Zealand (and indeed other countries) is to be less sceptical and more open to incorporating TEK into climate change preparedness and adaptation.

PACIFIC RESPONSES

Using detailed examples from several areas of the island Pacific, the adaptation measures discussed here demonstrate how local knowledge works in long-held beliefs and practice as well as contemporary science. The PICs, using examples of land building, relocation, agricultural practices, food storage and preservation, as well as a deep traditional understanding of changing weather patterns and their impacts on sources of sustenance and the challenges of contemporary settlements are deeply committed to a 'climate resilient future', including a Pacific of green blue economies.[1]

With the disappointing failure of the PIF in Port Moresby in September 2015 to endorse the less than 1.5 degrees of warming target, ahead of the Paris UN Climate Change Summit later that year, and the recent cuts by the United States to the Global Climate Fund, it is clear that the Pacific nations will need to not only become even more outspoken and vocal in their responses to the impacts of climate change, but also more proactive. They may not always respond in one voice, due to the diversity of the nations, but understanding of the impacts of climate change is widespread.

Pacific countries are considered to be particularly vulnerable to climate change, but what is also clear is that they have many ways of dealing with what now appears to be inevitable. Traditional practices, such as those involving disaster preparation and the building of artificial islands, as well as global lobbying and new forms of adaptation, will become not only more necessary but also increasingly practiced. The responses are not merely 'knee jerk' reactions. Local scientists, working with communities, will continue to utilize existing local solutions. At the regional level, new forms of governance are also emerging. There is some division at this level with Australia and New Zealand supporting only limited action on Climate Change and facing strong verbal censure from some leaders, as well as from the PIDF (see Chap. 2). There have even been claims 'that a Pacific Islands Forum with Australia and New Zealand as members ... hampers the ability of the Pacific island states to defend their interests, and in the case of climate change policy, their very survival'[2,3] but perhaps now is the time to make the exchange of expertise more even with the island Pacific perhaps teaching Aotearoa New Zealand (as well as other donor partners) how to utilize existing local and community knowledge. There is now no choice. 'Irreversible loss and damage ... goes beyond adaptation and is already a reality for Pacific Small Island Developing States'.[4]

It is increasingly clear that the PICs (excluding Aotearoa New Zealand) are united about the impacts of climate change. They have traditional ways of coping which can be models for other parts of the world, but the scale, cost and magnitude may well defeat some islands and preparation for other means of adaptation, such as relocation, are already underway. So too will the countries continue to contribute to global research and policy to 'turn the tide'. Although it may be too late for some low-lying islands, there is active engagement in seeking actions aimed at 'safeguarding bio-diversity and ecosystems; ensuring food, water and energy security; and supporting future socio-economic development by becoming climate resilient'.[5] The actions sought may well include the construction of appro-priately developed and cost effective artificial islands. Whatever is devel-oped must involve the agreement of Pacific Islanders themselves.

In Aotearoa New Zealand there is also capability and awareness around adaptation for sea-level rise, coastal storm surges and flooding as the examples cited in Carter's companion book attest. There are also some uniquely Aotearoa New Zealand challenges such as those that arise from New Zealand's reliance on agriculture and industrial activities, something that is not such a large part of the island Pacific countries included in this book, whose economies could be termed more 'blue' than green, given their large ocean areas.

THINKING OUTSIDE THE SQUARE: WHAT ARE THE PACIFIC ISLANDS DOING?

Blue Economies

In August 2017 PIDF hosted the Blue Economy Conference. Held in Suva Fiji and chaired by Solomon Islands, the conference was a call for action on key emerging issues challenging sustainable management and conservation of the Pacific Ocean and its resources. It involved not only government and regional body representatives and policy makers, but also civil society and private sector representatives, academics, researchers and scientists to focus on the latest innovations, lessons learned, best practices and private-public partnerships to influence policies for sustainable management and conservation of ocean resources. The overall goal of the conference, was to work towards the implementation of United Nations Sustainable Development Goal 14 (SDG) which emphasizes conservation and sustainable use of the oceans, seas and marine resources. This is not the first time the PIDF has hosted such a conference. The first was in 2016, and the fact that a wide range of people participate demonstrates recognition of the value of listening to and respecting the value of local knowledge. Indeed, recommendation two was powerful in directly recognizing and supporting cultural practices in the statement, 'A sustainable Blue Economy is one in which the use of our oceans today enhances rather than undermines natural capital and does not compromise the ability of future generations to generate cultural, social and economic wealth'.[6] The participants were in agreement that healthy oceans are fundamental, and that the solutions need to be 'locally grown'.[7]

The Pacific countries are also very strongly supportive of blue carbon initiatives, recognizing that coastal and ocean environments are capable of high carbon sequestration and so need to be preserved for overall environmental health, but also play a significant role in adaptation including acting as effective buffers against storm surges. Coastal blue carbon ecosystems occupy less than 1 per cent of the seafloor, yet sequester almost half the ocean's stored carbon. They store carbon 30–50 times faster than tropical rainforests, and can store it for very long periods[8] but few understand the value of seagrass. SPREP has suggested that a mapping exercise of Pacific seagrass be carried out, and it would also be useful to know what traditional knowledge there is around this important resource which is really a 'hidden' resource of Carbon Capture and Storage (CCS). Blue carbon sinks exist in the biomass and sediments of habitats such as mangroves, sea grass beds, tidal marshes, and other marine and coastal vegetated ecosystems'.[9] When these areas are degraded or destroyed, they release the carbon into the atmosphere. Carter refers in her book to the fact that blue carbon areas have not been included into carbon trading schemes, despite these ecosystems contributing significantly to atmospheric greenhouse gases. She also notes the potential economic benefits for indigenous and local communities in good management of these ecosystems.

PICs already have underway several attempts to capture and store carbon, such as the Sovi Basin in Fiji, and in Vanuatu. Given the opening Statement to COP16 on Blue Carbon solutions to Climate Change in 2010, made by a number of NGO representatives (including from Pacific and Australia and New Zealand)[10] which called for the establishment of a Blue Carbon Fund, it seems wise for Aotearoa New Zealand to revisit, take notice of and emulate CCS attempts in the Pacific.

Sovi Basin

One example for Aotearoa New Zealand and other nations to consider is that of Sovi in Fiji. For the past three decades, there has been interest shown in the future of this area, a forested and diverse ecosystem of around 19,600 hectares (50,000 acres) on the large island of Viti Levu. This quite remote area has been considered at various times as an important carbon sink, and most recently as an area worth protecting under the UNESCO World Heritage cultural category. In a partnership between Conservation International and Fiji Water, a bottled water company based in Fiji, there is an initiative to conserve and protect the area as one of the few remaining virgin tropical forests in inland Fiji. In response to Fiji's commitments to the UNFCCC, the Sovi Basin carbon sequestration programme aimed to promote sustainable management, and promote and cooperate in the conservation and enhancement of sinks and reservoirs of all greenhouse gases. There was also the addition of the Sovi Basin Trust Fund (SBTF), which is supported by donations from Conservation International. This fund makes annual disbursements that not only offset the cash value of logging pay outs to local landowners, but also pay for land leases and create jobs. Such an approach is considered to be innovative and unusual in the Fiji context, but issues of land tenure continue to raise problems as landowners of the resource and the partners cannot come to an agreement over lease arrangements.[11] The fact that Sovi Basin remains on the tentative list for UNESCO, and has not become a successful example of carbon sink sequestration is a pity, but does point to the complexity and time required for PICs in resolving issues surrounding land tenure.

More recent plans such as for Vanuatu to realize the potential of carbon sinks[12] are also underway following a scoping study by the Commonwealth Secretariat. This work aims to showcase what is possible in blue carbon, but cautions that it is not only to do with financial reward, but is instead looking towards community and ecosystem benefits, crucial for livelihoods in smaller countries, particularly those with large areas of mangrove. Vanuatu had examined issues of carbon sinks before, and organizations, including governments, were aware of the risky nature of carbon trading, but in the past three years, working closely with other projects, such as REDD+ is recruiting staff and establishing mangrove projects in what may well become a positive example in the Pacific towards managing coastal (blue) carbon sinks.

Not all Pacific countries are as engaged with REDD+.[13] There are many more projects that could be emulated by other suitable nations with forests of significance and that, significantly, are being taken up across the Pacific Islands. Carter, in her companion book, notes new forms of technology as well as the adaptation of early and traditional technologies that could be enhanced to improve the livelihoods of Pacific Island nations. 3D printing of coral reefs is no longer a fantasy, and may well be the saviour for the world's rapidly dying coral reefs, and the adaptation of floating houses to enable people to continue living in areas that flood is already underway in London and Amsterdam, but is of course long known in several areas of the Pacific. The sharing of new technologies that can enhance the old, and the adaptation of the earlier technologies that can protect the new, may well provide an answer to the significant impacts of climate change.

Other adaptations, such as a return to large sailing vessels, capable of carrying large loads of cargo and travelling long distances are now not out of the question. In a world where sea transport produces as many greenhouse gasses as flight[14] the re-emergence of such vessels, enhanced with solar power and fibre-glass hulls are not at all a fantasy. Already these are being used to transport goods across parts of the Pacific, such as between Papua New Guinea and Solomon Islands, and could ultimately solve the problems of irregular inter-island shipping within the Pacific Islands, as well as saving on fuel imports and the production of GHGs.

What the energy behind these projects as well as other adaptation examples discussed in earlier chapters demonstrate, is that time, scepticism and failure to act are not options for the smaller PICs. Even though the emissions of Pacific countries are negligible, the human energy that has already gone into the return to 'soft' sailing ships, and the Sovi and Vanuatu projects may still act as lessons for many parts of the world, and perhaps most especially to Aotearoa New Zealand as a major donor to and close partner of the Pacific.

The core argument of this book is that Pacific Island nations have a great deal of knowledge of how to manage their environments and respond to the impacts of climate change. Listening to local communities and respecting knowledge that already exists is frequently overlooked by donor countries and others purporting to assist. By taking some of the lessons from the Pacific and applying and adapting them to other nations, global responses to climate change could be more sustainable. The relationship with Aotearoa New Zealand is a main focus of this book because of its long and close ties with many Pacific nations, but it does not have to be the only country that can learn from Pacific countries. As demonstrated here, there are many lessons to be learned and shared.

PICs, as demonstrated throughout this book, are fighting to be heard and to act on the challenge from climate change. The countries have demonstrated through their many efforts at adaptation and sheer determination that there are ways of coping that involve significant cross-party agreements, major legislation and virtually no scepticism. The impacts of climate change, exacerbated by the behaviour of humans throughout the world, are now affecting not only the low-lying atolls of the Pacific, but everywhere, including their neighbours of Aotearoa New Zealand and Australia. The issue is not one for PICs alone, but requires their larger industrialized neighbours, to step up and take responsibility for developing and actioning effective mitigation and adaptation policies. PICs have shown the way for the two key climate change threats identified here: sea-level rise and flooding. They are also thinking forward and inter-generationally by investing time and money into innovative new adaptations that often build on tradition. These are important lessons that many other countries could learn from. As said in Samoa, climate change is not everyone else's problem and we should all take heed of the following advice:

Climate change is a global problem with grave implications: environmental, social, economic, political, and for the distribution of goods. It represents one of the principal challenges facing humanity in our day. Its worst impact will probably be felt by developing countries in coming decades. Many of the poor live in areas particularly affected by phenomena related to warming, and their means of subsistence are largely dependent on natural reserves and ecosystemic services such as agriculture, fishing and forestry.

The climate is a common good, belonging to all and meant for all.[15]

NOTES

1. The Third Pacific Islands Development Forum Leaders' Summit, held in Suva in September 2015, had as its overarching theme 'Building Climate Resilient Green Blue Pacific Economies' and in 2017 PIDF signed a Memorandum of Understanding with the Global Green Growth Institute (GGGI).
2. G. Fry, http://devpolicy.org/pacific-climate-diplomacy-and-the-future-relevance-of-the-pacific-islands-forum-20150904/.
3. It is fully understood that despite the absence of Australia and New Zealand at the PIDF meetings, other powers, such as China, were welcomed.
4. Pacific Island Development Forum Secretariat, 'Suva Declaration on Climate Change', Suva, Fiji, 4 September 2015.
5. PIDF Summit Programme, PIDF Secretariat, Suva, September 2015.
6. http://pacificidf.org/wp-content/uploads/2017/08/Conference-Summary_3-September.pdf.
7. Recommendation 6, http://pacificidf.org/wp-content/uploads/2017/08/Conference-Summary_3-September.pdf.
8. http://www.sprep.org/climate-change/international-partnership-for-blue-carbon.
9. Vierros (2013).
10. Blue Climate Coalition, 'Blue Carbon Solutions for Climate Change' Open statement to the delegates of COP16. *blueclimatesolutions.org*, 30 November 2010.
11. Final Report National Assessment, Climate Change United Nations Framework Convention On Climate Change (UNFCCC), Department of Environment, NCSA Project Steering Committee, NCSA Project and UNDP Fiji, October 2008.
12. Laffoley (2013).
13. Reducing Emissions from Deforestation and Forest Degradation is of more benefit to higher islands with forest resources. The divergence in views over REDD+ has led to some discontent among smaller nations.

The inclusion of mangrove areas in REDD+ will go some of the way to mitigate divisions, but also demonstrates that there cannot be a single Pacific voice on all issues.

14. Maritime transport emits around 1000 million tonnes of CO_2 annually and is responsible for about 2.5 per cent of global greenhouse gas emissions. Depending on future economic and energy developments, marine emissions shall increase between 50 per cent and 250 per cent by 2050 https://ec.europa.eu/clima/policies/eccp_en.

15. Encyclical Letter Laudato of the Holy Father Francis on Care for our Common Home Libreria Editrice Vaticana, 24 May 2015.

GLOSSARY

Buli Head man (Fijian)
Bus Bush (Solomon Islands Pijin)
Isa That's a pity (exclamation of regret in Fijian)
Kerekere Borrowing (Fijian)
Lali Drum (Fijian)
Sol wara Salt water (Solomon Islands Pijin)
Tauvu Relationship between people (lit. worshipper of the same god) (Fijian)
Vina'a va'a levu saroga Thank you very much (in Fijian Buan dialect) (ch. Vinaka vaka levu which is the standard Bauan dialect)

© The Author(s) 2018   95
J. Bryant-Tokalau, *Indigenous Pacific Approaches to Climate Change*, Palgrave Studies in Disaster Anthropology,
https://doi.org/10.1007/978-3-319-78399-4

References

Allen, K. M. (2006, March). Community-Based Disaster Preparedness and Climate Adaptation: Local Capacity-Building in the Philippines. *Disasters: Special Issue: Climate Change and Disasters, 30*(1), 81–101. https://doi.org/10.1111/j.1467-9523.2006.00308.x.

Asian Development Bank. (2008). *Kiribati: Preparing the Outer Island Growth Centers Project – Phase 2.* The Proposed Water Supply and Sanitation Project Executive Report (Volume 1) Project Number: TA 4456 – KIR. Prepared by Paul Jones, Sinclair Knight Merz, Melbourne, Australia. For Ministry of Finance & Economic Development (MFED) & Ministry of Line & Phoenix Islands Development (MLPID).

Asian Development Bank. (2010). *Responding to Climate Change in the Pacific: Moving from Strategy to Action.* Pacific Studies Series. Manila: ADB.

Australian Bureau of Meteorology & CSIRO. (2011). Climate Change in the Pacific: Scientific Assessment and New Research. Volume 2: Country Reports.

Ayres, W. S. (1983). Archaeology at Nan Madol, Pohnpei. *Bulletin of the Indo-Pacific Prehistory Association.* Retrieved August 19, 2010, from http://ejournal.anu.edu.au/index.php/bippa/article/viewFile/480/469.

Barnett, J., & Campbell, J. (2010). *Climate Change and Small Island States: Power, Knowledge and the South Pacific.* London: Earthscan.

Barnett, J., & Waters, E. (2016). Rethinking the Vulnerability of Small Island States: Climate Change and Development in the Pacific Islands. In J. Grugel & D. Hammett (Eds.), *The Palgrave Macmillan Handbook of International Development* (pp. 731–748). London: Palgrave Macmillan.

© The Author(s) 2018
J. Bryant-Tokalau, *Indigenous Pacific Approaches to Climate Change*, Palgrave Studies in Disaster Anthropology,
https://doi.org/10.1007/978-3-319-78399-4

Bayliss-Smith, T. P. (1977). Hurricane Val in North Lakeba: The View from 1975. In R. McLean, T. P. Bayliss-Smith, M. Brookfield, & J. R. Campbell (Eds.), *The Hurricane Hazard: Natural Disaster and Small Population* (pp. 65–98). Population and Environment Project in the Eastern Islands of Fiji. Island Reports 1. Man and the Biosphere Programme Project 7: Rational Use of Island Ecosystems, Canberra.

Bell, J., Taylor, M., Amos, M., & Andrew, N. (2016). *Climate Change and Pacific Island Food Systems*. The Future of Food, Farming and Fishing in the Pacific Islands Under a Changing Climate. CCAFS and CTA, Copenhagen, Denmark and Wageningen, The Netherlands.

Bernard, K., & Cook, S. (2014). Luxury Tourism Investment and Flood-Risk: Case Study on Unsustainable Development in Denarau Island Resort in Fiji. *International Journal of Disaster Risk Reduction*. https://doi.org/10.1016/j.ijdrr.2014.09.002.

Bettencourt, S., Croad, R., Freeman, P., Hay, J., Jones, R., King, P., Lal, P., Mearns, A., Miller, G., Pswarayl-Riddihough, I., Simpson, A., Teuataloo, N., Trotz, U., & Van Aalst, M. (2006). *Not If But When: Adapting to Natural Hazards in the Pacific Island Regions*. Washington, DC: Policy Note, World Bank.

Bhagwan, J. (2014, April 6). Spirituality and the Environment. *The Fiji Times Online*. Retrieved April 6, 2015. www.fijitimes.com/story.aspx?.idxxx

Biribo, N., & Woodroffe, C. (2013). Historical Area and Shoreline Change of Reef Islands Around Tarawa Atoll, Kiribati. *Sustainability Science, 8*(3), 345–362.

Brookfield, M. (1977). Hurricane Val and Its Aftermath: Report on an Inquiry Among the People of Lakeba in 1976. In R. McLean, T. P. Bayliss-Smith, M. Brookfield, & J. R. Campbell (Eds.), *The Hurricane Hazard: Natural Disaster and Small Population* (pp. 99–147). Population and Environment Project in the Eastern Islands of Fiji. Island Reports 1. Man and the Biosphere Programme Project 7: Rational Use of Island Ecosystems, Canberra.

Bryant, J. (1993). *Urban Poverty and the Environment in the South Pacific*. Armidale: University of New England.

Bryant-Tokalau, J. (1994). Our Changing Islands: Pacific Urban Environments. *The Courier, 144,* 80–82.

Bryant-Tokalau, J. (2008). From Summitry to Panarchy: Issues of Global, Regional and Indigenous Environmental Governance in the Pacific. *Borderlands e-Journal: New Spaces in the Humanities, 7*(3). Certainty in the Coming Community J. Mummery and V. Devadas (eds.).

Bryant-Tokalau, J. (2010). Living in the Qoliqoli: Urban Squatting on the Fiji Foreshore. *Pacific Studies, 33*(1), 1–20.

Bryant-Tokalau, J. (2011). Artificial and Recycled Islands in the Pacific: Myths and Mythology of "Plastic Fantastic". *The Journal of the Polynesian Society, 120*(1), 71–86.

Bryant-Tokalau, J. (2012). Twenty Years On: Poverty and Hardship in Urban Fiji. *Bijdragen tot de Taal, Land en-Volkenkunde, 168*(2/3), 195–218.

Bryant-Tokalau, J. (2013). The Changing Face of the Urban Pacific. In D. Dussy & E. Wittersheim (Eds.), *Villes Invisibles: Anthropologie Urbaine dans le Pacifique L'Harmattan* (pp. 45–74). Paris: Collection "Cahiers du Pacifique Sud Contemporain".

Bryant-Tokalau, J. (2014a). Urban Squatters and the Poor in Fiji: Issues of Land and Investment in Coastal Areas. *Asia Pacific Viewpoint, 55*(1), 54–66. ISSN 1360-7456.

Bryant-Tokalau, J. (2014b, December 3). *Indigenous Responses to Environmental Challenges: Artificial Islands and the Challenges of Relocation.* Paper presented session, 'Climate Change, Disasters and Pacific Agency', *Pacific History Conference, Lalan Chalan Tala Ara,* Taipei, Taiwan.

Bryant-Tokalau, J. (2016a). Human & Environmental Security: What the Pacific Can Teach NZ & Australia About Climate Change. Macmillan Brown Center for Pacific Studies: Pacific Policy Brief 2016/2. ISSN 1172-3416.

Bryant-Tokalau, J. (2016b). *Climate Change and Health in the Pacific Region: How People Adapt and What Should be NZ/Aotearoa's Involvement.* Presentation to University of Otago Summer School Climate Change and Health, Wellington.

Bryant-Tokalau, J. (2016c). Community Responses to Floods in Fiji: Lessons Learned. In J. Calabrese (Ed.), Chapter for Middle East Asia Project (MAP) Series on '*Humanitarian Assistance and Disaster Response: Rising to the Challenge*'. Washington, DC: Middle East Institute. http://www.mei.edu/content/map/community-responses-floods-fiji-lessons-learned.

Bryant-Tokalau, J., & Campbell, J. (2014). Coping with Floods in Urban Fiji: Responses and Resilience of the Poor. In E. Jurriens (Ed.), *Disaster Relief in the Asia Pacific Region: Capacity Building and Community Resilience* (pp. 132–146). Oxford: Routledge.

Cajete, G. (2000). *Native Science Natural Laws of Interdependence.* Santa Fe: Clear Light Publishers.

Campbell, J. R. (2006). *Traditional Disaster Reduction in Pacific Island Communities.* GNS Science Report 2006/038. Institute of Geological and Nuclear Sciences, Lower Hutt, New Zealand.

Campbell, J. R. (2010). An Overview of Natural Hazard Planning in the Pacific Island Region. *The Australasian Journal of Disaster and Trauma Studies,* 1. Online. Retrieved April 13, 2013, from http://trauma.massey.ac.nz/issues/2010-1/campbell.htm.

Campbell, J. (2014). Climate-Change Migration in the Pacific. *The Contemporary Pacific, 26*(1), 1–28.

Campbell, J., & Bedford, R. (2014). Migration and Climate Change in Oceania. *Global Migration Issues, 2,* 177–204.

Carew-Reid, J. (1989). *Environment, Aid and Regionalism in the South Pacific.* Pacific Research Monograph No. 22. Canberra: National Centre for Development Studies, ANU.

Caritas. (2014). *Small Yet Strong: Voices from Oceania on the Environment.* Wellington: Caritas Aotearoa.

Carter, G. (2015). Establishing a Pacific Voice in the Climate Change Negotiations. In G. Fry & S. Tarte (Eds.), *The New Pacific Diplomacy* (pp. 205–220). Canberra: ANU Press.

Chandra, R. (1990). Patterns and Processes of Urbanisation in Fiji. In R. Chandra & J. Bryant (Eds.), *Population of Fiji* (pp. 157–179). Population Monograph No. 1. Noumea: South Pacific Commission.

Chandra, A., & Dalton, J. A. (2011, May 12). *Managing Watersheds for Urban Resilience.* Partnership for Environment and Disaster Risk Reduction (PEDRR). Policy Brief Presented at the Global Platform for Disaster Risk Reduction, Roundtable on "Managing Watersheds for Urban Resilience", Geneva, Switzerland.

Clarke, W. C. (1990). Learning from the Past: Traditional Knowledge and Sustainable Development. *The Contemporary Pacific, 2*(2), 233–253.

Connell, J. (2010). Pacific Islands in the Global Economy: Paradoxes of Migration and Culture. *Singapore Journal of Tropical Geography, 31*, 115–129.

Connell, J. (2012). Population Resettlement in the Pacific: Lessons from a Hazardous History? *Australian Geographer, 43*(2), 127–142.

Connell, J. (2013). Soothing Breezes? Island Perspectives on Climate Change and Migration. *Australian Geographer, 44*(4), 465–480.

Connell, J. (2016). Last Days in the Carteret Islands? Climate Change, Livelihoods and Migration on Coral Atolls. *Asian Pacific Viewpoint, 57*(1), 3–15.

Connell, J., & Lea, J. (2002). *Urbanisation in the Island Pacific: Towards Sustainable Development.* London: Routledge.

D'Arcy, P. (2006). *The People of the Sea: Environment, Identity, and History in Oceania.* Honolulu: University of Hawai'i Press.

Donner, S. D. (2015). Climate Change: Fantasy Island. *Scientific American, 24*(1), 50–57.

Eco-Bishops. (2015, February 26). Learning from Indigenous Peoples. #pisky #ecobishops#anglican. www.bishopdavid.net/2015/02.

Fry, G. (2015). Pacific Climate Diplomacy and the Future Relevance of the Pacific Islands Forum. Devpolicy Blog. Retrieved November 18, 2015, from http://devpolicy.org/pacific-climate-diplomacy-and-the-future-relevance-of-the-pacific-islands-forum-20150904/.

Fry, G., & Tarte, S. (Eds.). (2016). *The New Pacific Diplomacy.* Canberra: ANU Press, Pacific Series.

Gallen, S. L. (2016). Micronesian Sub-Regional Diplomacy. In G. Fry & S. Tarte (Eds.), *The New Pacific Diplomacy* (pp. 175–188). Canberra: ANU Press, Pacific Series.

Garnaut, R. (2008). *The Garnaut Climate Change Review: Final Report.* Cambridge: Cambridge University Press.

Gegeo, D. W. (1998). Indigenous Knowledge and Empowerment: Rural Development Examined from Within. *The Contemporary Pacific, 10*(2), 289–315.

Gero, A., Fletcher, S., Rumsey, M., Thiessen, J., Kuruppu, N., Buchan, J., Daly, J., & Willetts, J. (2015). Disasters and Climate Change in the Pacific: Adaptive Capacity of Humanitarian Response Organizations. *Climate and Development, 7*(1), 35–46. https://doi.org/10.1080/17565529.2014.899888. https://doi.org/10.1080/17565529.2014.899888.

Glacken, C. J. (1967). *Traces on the Rhodian Shore: Nature and Culture in Western Thought from Ancient Times to the End of the Eighteenth Century.* Berkeley: University of California Press.

Goulding, N. (2016). Marshalling a Pacific Response to Climate Change. In G. Fry & S. Tarte (Eds.), *The New Pacific Diplomacy* (pp. 191–204). Canberra: ANU Press, Pacific Series.

Government of Fiji. (2012). Republic of Fiji. 'The National Housing Policy: Affordable and Decent Housing for All', Suva.

Government of Kiribati. (2013). *Second Communication Under the United Nations Framework Convention on Climate Change.* Environment & Conservation Division, with Assistance of Climate Change Study Team Ministry of Environment, Lands and Agricultural Development.

Granderson, A. A. (2017). The Role of Traditional Knowledge in Building Adaptive Capacity for Climate Change: Perspectives from Vanuatu. *Weather, Climate and Society.* https://doi.org/10.1175/WCAS-D-16-0094.1.

Guo, P.-Y. (2001). *Landscape, History and Migration Among the Langalanga, Solomon Islands.* PhD dissertation in Anthropology, University of Pittsburgh.

Haberkorn, G. (2008). Pacific Islands' Population and Development: Facts, Fictions and Follies. *New Zealand Population Review, 33/34,* 95–127.

Hau'ofa, E. (1994, Spring). Our Sea of Islands. *The Contemporary Pacific, 6*(1), 148–161.

Heijmans, A. (2004). From Vulnerability to Empowerment. In G. Bankoff, G. Ferks, & D. Hilhorst (Eds.), *Mapping Vulnerability: Disasters, Development and People* (pp. 115–127). London: Earthscan.

Hermann, E., & Kempf, W. (2017). Climate Change and the Imagining of Migration: Emerging Discourses on Kiribati's Land Purchase in Fiji. *The Contemporary Pacific, 29*(2), 232–263.

Hviding, E. (1998). Contextual Flexibility: Present Status and Future of Customary Marine Tenure in Solomon Islands. *Ocean and Coastal Management, 40,* 253–269.

Hviding, E. (2016). Europe and the Pacific: Engaging Anthropology in EU Policy Making and Development Cooperation. In H. Hviding (Ed.), *Engaged Anthropology* (pp. 147–166). Berlin: Springer.

Iati, I. (2010). China and Samoa. In T. Wesley-Smith & E. A. Porter (Eds.), *China in Oceania: Reshaping the Pacific?* (pp. 151–163). New York: Berghahn Books.

Ikeda, D. (2014). *Value Creation for Global Change: Building Resilient and Sustainable Societies.* Tokyo: Peace Proposal, Soka Gakkai International.

Ivens, W. G. (1930). *The Island Builders of the Pacific: How and Why the People of Mala Construct Their Artificial Islands, the Antiquity & Doubtful Origin of the Practice, with a Description of the Social Organisation, Magic, & Religion of the Inhabitants.* London: Seeley, Service & Co.

Kardol, R. (1999). *Proposed Inhabited Artificial Islands in International Waters: International Law Analysis in Regards to Resource Use, Law of the Sea and Norms of Self-Determination and State Recognition.* Master's thesis at Universiteit van Amsterdam, The Netherlands. http://seasteading.org/seastead.org/localres/misc-articles/kardol1999.html.

Keen, M., & McNeill, A. (2016). *After the Floods: Urban Displacement, Lessons from Solomon Islands.* SSGM Brief, ANU 2016/13.

Kelman, I. (2010, August 9). https://groups.google.com/forum/#!forum/sicri-news-list.

Kelman, I., Lewis, J., Gaillard, J. C., & Mercer, J. (2011). Participatory Action Research for Dealing with Disasters on Islands. *Island Studies Journal, 6*(1), 59–86.

Korauaba, T. (2012). *Media and the Politics of Climate Change in Kiribati: A Case Study on Journalism in a "Disappearing Nation".* Master of Communication dissertation, Auckland University of Technology.

Laffoley, D. d'A. (2013). *The Management of Coastal Carbon Sinks in Vanuatu: Realising the Potential.* A Report to the Government of Vanuatu, Commonwealth Secretariat, London.

Löfgren, O. (2007). Island Magic and the Making of a Transnational Region. *Geographical Review, 97*(2), 244–260.

MacLellan, N. (2015a). *Look Out Australia the Pacific Mood Is Shifting.* http://www.lowyinterpreter.org/post/2015/02/04/australia-pacific-mood-shifting.

MacLellan, N. (2015b). Preparing for Cyclones Reuben, Solo, Tuni, Ula … and Beyond. Retrieved March 20, 2015, from http://insidestory.org.au/preparing-for-cyclones-reuben-solo-tuni-ula-and-beyond.

Manoa, F. (2016). The New Pacific Diplomacy at the United Nations: The Rise of the PSIDS. In G. Fry & S. Tarte (Eds.), *The New Pacific Diplomacy* (pp. 89–98). Canberra: ANU Press, Pacific Series.

McLean, R. (1977). The Hurricane Hazard in the Eastern Islands of Fiji: An Historical Analysis. In R. McLean, T. P. Bayliss-Smith, M. Brookfield, & J. Campbell (Eds.), *The Hurricane Hazard: Natural Disaster and Small Populations, Population and Environment Project in the Eastern Islands of Fiji, Man and the Biosphere Programme Project 7: Ecology and Rational Use of Island Ecosystems.* Canberra: ANU Development Studies Centre.

McMillen, H. L., Ticktin, T., Friedlander, A., Jupiter, S. D., Thaman, R., Campbell, J., Veitayaki, J., Giambelluca, T., Nihmei, S., Rupeni, E., Apis-Overhoff, L., Aalbersberg, W., & Orcherton, D. F. (2014). Small Islands, Valuable Insights: Systems of Customary Resource Use and Resilience to Climate Change in the Pacific. *Ecology and Society, 19*(4), 44. https://doi.org/10.5751/ES-06937-190444.

Mohanty, M. (2006). Squatters, Vulnerability and Adaptability of Urban Poor in a Small Island Developing State: The Context of Fiji Islands. Paper presented to IGU Commission on Population and Vulnerability and Asia Pacific Migration Research Network (APMRN), IGU 2006 Conference, Brisbane, 3–7 July. Online. Retrieved January 18, 2012, from www.hss.adelaide.edu.au/socialsciences/igu/documents/manoranjan_mohanty.pdf.

Morrison, K. (2017). The Role of Traditional Knowledge to Frame Understanding of Migration as Adaptation to the "Slow Disaster" of Sea Level Rise in the Pacific. In K. Sudmeier-Rieux et al. (Eds.), *Identifying Emerging Issues in Disaster Risk Reduction, Migration, Climate Change and Sustainable Development* (pp. 249–266). Cham: Springer International Publishing. https://doi.org/10.1007/978-3-319-33880-4_15.

Morrow, G., & Bowen, K. (2014). Accounting for Health in Climate Change Policies: A Case Study of Fiji. *Global Health Action, 7*. Retrieved May 21, 2014, from http://www.globalhealthaction.net/index.php/gha/article/view/23550/html.

Nunn, P. D. (1991). *Keimami sa vakila na liga ni Kalou* (Feeling the Hand of God): Human and Nonhuman Impacts on Pacific Island Environments. Occasional Papers of the East-West Environment and Policy Institute. Paper No. 13. Honolulu.

Nunn, P. D. (1994). *Oceanic Islands.* Oxford: Blackwell.

Nunn, P. D. (1999). Geomorphology. In M. Rapoport (Ed.), *The Pacific Islands: Environment and Society* (pp. 43–55). Honolulu: The Bess Press.

Nunn, P. D. (2009a). Responses to the Challenges of Climate Change in the Pacific Islands: Management and Technological Imperatives. *Climate Research, 40,* 211–231. Inter-Research. www.int-res.com.

Nunn, P. D. (2009b). *Vanished Islands and Hidden Continents of the Pacific.* Honolulu: University of Hawaii Press.

Nunn, P. D. (2017, May 17). Conversation, *Sidelining God: Why Secular Climate Projects in the Pacific Islands Are Failing* Relation Conversation. research.usc.edu.au, https://theconversation.com/sidelining-god-why-secular-climate-projects-in-the-pacific-islands-are-failing-77623.

Nunn, P. D., & Britton, J. M. R. (2001). Human-Environment Relationships in the Pacific Islands Around A.D. 1300. *Environment and History, 7,* 3–22.

Nunn, P. D., Mulgrew, K., Scott-Parker, B., Hine, D. W. Marks, A. D. G., Mahar, D., & Maebuta, J. (2016a). Spirituality and Attitudes Towards Nature in the Pacific Islands: Insights for Enabling Climate-Change Adaptation. *Climatic Change*. https://doi.org/10.1007/s10584-016-1646-9.

Nunn, P. D., Runman, J., Falanruw, M., & Kumar, R. (2016b, April 19). Culturally Grounded Responses to Coastal Change on Islands in the Federated States of Micronesia, North West Pacific Ocean. *Regional Environmental Change*. Online. https://doi.org/10.1007/S10113-016-0950-2.

Pacific Institute of Public Policy. (2012). Climate Security: A Holistic Approach to Climate Change, Security and Development. Discussion Paper 23, Port Vila.

Pala, C. (2014, August 21). The Nation that Bought a Back-Up Property. *The Atlantic*.

Parsonson, G. (1966). Artificial Islands in Melanesia: The Role of Malaria in the Settlement of the Southwest Pacific. *New Zealand Geographer, 22*(1), 1–21.

Powles, M. (2010). Challenges, Opportunities and the Case for Engagement. In T. Wesley-Smith & E. A. Porter (Eds.), *China in Oceania: Reshaping the Pacific?* (pp. 67–84). New York: Berghahn Books.

Raj, R. (2004). *Integrated Flood Management. Case Study 1.* Fiji Islands: Flood Management – Rewa River Basin. Edited by Technical Support Unit World Meteorological Organization/Global Water Partnership. The Associated Programme on Flood Management.

Rakova, U. (2017, July). The Carterets: What is it Like to be driven from your home by something that is out of your control. 'Tulele Peisa'. Presentation to University of Otago Foreign Policy School.

Rapoport, M. (Ed.). (1999). *The Pacific Islands: Environment and Society.* Honolulu: The Bess Press.

Ravuvu, A. (1988). *Development or Dependence: The Pattern of Change in a Fijian Village.* Suva: Institute of Pacific Studies, The University of the South Pacific.

Rudiak-Gould, P. (2013). *Climate Change and Tradition in a Small Island State: The Rising Tide.* Oxford: Routledge Studies in Anthropology.

Secretariat of the Pacific Community. (2015). *Pocket Statistical Summary.*

Smith, R. (2013). Should They Stay or Should They Go? A Discourse Analysis of Factors Influencing Relocation Decisions Among the Outer Islands of Tuvalu and Kiribati. *Journal of New Zealand and Pacific Studies, 1*(1), 23–39.

SOPAC. (2009). *Relationship Between Natural Disasters and Poverty: A Fiji Case Study.* SOPAC Miscellaneous Report 678. Prepared for UN International Strategy for Disaster Reduction Secretariat's 2009 Global Assessment Report on Disaster Reduction.

SOPAC. (2012). *Enhancing Coastal Resilience in Kiribati: An Integrated Approach to Addressing Beach Mining in Urban South Tarawa.* Pacific Geoscience Commission, Suva. (Video Recording Produced with EU and Government of Kiribati).

South Pacific Regional Environment Programme. (1992). The Pacific Way: Pacific Island Developing Countries' Report to the United Nations Conference on Environment & Development, Noumea.

South, R., & Veitayaki, J. (1999). *Global Initiatives in the South Pacific: Regional Approaches to Workable Arrangements.* Asia Pacific School of Economics and Management Studies Online. Asia Pacific Press, Australian National University, Canberra. Retrieved from http://ncdsnet.anu.edu.au.

Storey, D. (2005). *Urban Governance in Pacific Island Countries: Advancing an Overdue Agenda.* SSGM Discussion Paper 2005/7. Canberra, Australian National University.

Tabe, T. (2014, December 3). *The First Encounter: Reconceptualizing the Relocation of the Gilbertese Settlers Form Atolls in Micronesia to High Islands in Melanesia.* Presentation to session, 'Climate Change, Disasters and Pacific Agency', *Pacific History Conference, Lalan Chalan Tala Ara,* Taipei, Taiwan.

Tabe, T. (2016). *Ngaira Kain Tari – We Are People of the Sea: A Study of the Gilbertese Resettlement to Solomon Islands.* PhD dissertation, University of Bergen.

Teaiwa, K. M. (2015). *Consuming Ocean Island: Stories of People and Phosphate from Banaba.* Bloomington: Indiana University Press.

Tomlinson, M. (2017). Try the Spirits: Power Encounters and Anti-Wonder in Christian Missions. https://doi.org/10.1080/20566093.2017.1351171.

Tong, A. (2016). 'Charting Its Own Course': A Paradigm Shift in Pacific Diplomacy. In G. Fry & S. Tarte (Eds.), *The New Pacific Diplomacy* (pp. 21–24). Canberra: ANU Press, Pacific Series.

Van Veldhuizen, M. (2014). *Regional Organisations and Climate Change Adaptation in Pacific Island Developing States: An Analysis of the Regional Institutional Framework for Climate Change Adaptation in the Pacific Small Island Developing States and Territories.* M.Sc. dissertation in Environmental Sciences, Policy and Management, Erasmus Mundus, EC.

Veitayaki, J. (1997). Traditional Marine Resource Management Practices Used in the Pacific Islands: An Agenda for Change. *Ocean and Coastal Management, 37*(1), 123–136.

Veitayaki, J., Waqalevu, V., Varea, R., & Rollings, N. (2017). Mangroves in Small Island Development States in the Pacific: An Overview of a Highly Important and Seriously Threatened Resource. In R. DasGupta & R. Shaw (Eds.), *Participatory Mangrove Management in a Changing Climate. Disaster Risk Reduction (Methods, Approaches and Practices).* Tokyo: Springer.

Vierros, M. (2013, October 10). Communities and Blue Carbon: The Role of Traditional Management Systems in Providing Benefits for Carbon Storage, Biodiversity Conservation and Livelihoods. *Springer Science+Business Media Dordrecht, Special Edition: Climate Change.* Retrieved 2015, from https://doi.org/10.1007/s10584-013-0920-3.

Walter, R. K., & Hamilton, R. J. (2014). A Cultural Landscape Approach to Community-Based Conservation in Solomon Islands. *Ecology and Society*, *19*(4), 41. https://doi.org/10.5751/ES-06646-190441.

Webb, A. P., & Kench, P. S. (2010). The Dynamic Response of Reef Islands to Sea-Level Rise: Evidence from Multi-Decadal Analysis of Island Change in the Central Pacific. *Global and Planetary Change*. https://doi.org/10.1016/j.gloplacha.2010.05.03.

World Bank. (2016). *Climate and Disaster Resilience*, 69pp. http://pubdocs.worldbank.org/en/720371469614841726/PACIFIC-POSSIBLE-Climate.pdf.

Wyett, K. (2013). Escaping a Rising Tide: Sea Level Rise and Migration. *Asia & the Pacific Policy Studies*, *1*(1), 175–185.

Yeo, S., & Blong, R. J. (2010). Fiji's Worst Natural Disaster: The 1931 Hurricane and Flood. *Disaster*, *34*(3), 657–683.

INDEX[1]

[1]Note: Page numbers followed by 'n' refer notes.

© The Author(s) 2018
J. Bryant-Tokalau, *Indigenous Pacific Approaches to Climate Change*, Palgrave Studies in Disaster Anthropology, https://doi.org/10.1007/978-3-319-78399-4